Geology of the cou [lead and Camelford]

The Trevose Head and Camelford district is an area of outstanding natural beauty, encompassing a range of landscapes which reflect the variety of underlying rocks and the erosive processes acting upon them. The brooding granite uplands of Bodmin Moor rise sharply from the surrounding country rocks in the eastern part of the district, to dominate the inland scene.

The treeless tableland flanking the moor, which extends westwards towards the majestic coastline, conceals isolated, deeply incised wooded valleys of great charm which are rarely discovered by visitors. Such valleys were formed when sea level fell during periods of glacial advance in northern Europe. Now that the climate has ameliorated, the sea level has risen to drown the lower reaches of the river system, and the sea has been carried deep into the countryside through the Camel Estuary. This ria is rapidly silting up to give mud flats and sand banks which teem with wildlife. The tranquillity of this landscape belies a dynamic and often violent earth history extending back some 400 million years.

This, the second edition of a memoir originally published in 1910, has been completely rewritten, and incorporates accounts of all the geological formations in the district, the granite, and the resources of the extractive industry.

The area records the development of the Trevone Basin in which some 6000 m of marine sediments accumulated. This basin represents an area of persistent downwarping, developed on the broad continental shelf extending southwards from the Caledonian Continent.

Such subsidence was achieved by continued movements on deep-seated fractures in the earth's crust. These faults also provided pathways allowing great quantities of magma to reach the surface producing thick, pillowed lava flows and tuffs.

After a history of accumulation of sediments, which probably continued for some 70 million years, the contents of the basin were squeezed out as a deformation front spread northwards across the region in Carboniferous times. Folding and thrusting with opposed directions of transport developed, to give a geological structure of great complexity. The last major event associated with this mountain building episode was the intrusion of the Bodmin Moor Granite. As the magma cooled, volatile-enriched liquids deposited metallic ores both within and adjacent to the slowly cooling body. At the same time, hot water circulating through the rock led to the local alteration of the granite and to the formation of kaolin-rich china clay.

Few younger deposits have survived to tell of the final chapter of the geological history of this district. More striking are the planated surfaces recording repeated erosional episodes which reduced a mountain chain to the present landscape.

The complexities of the geology of the district have led to contentious interpretations which have featured in regional geological modelling. In this memoir, no attempt has been made to give a detailed evaluation of these models; rather it is hoped that the data recorded, will help to constrain the issues.

Cover photograph

A northerly view from near Bedruthan Steps (SW 8484 6937) towards Park Head. Red Cove Islan central foreground, Samaritan Island and Queen Bess Rock are within the Bedruthan Formation. Diggory's Island beyond, and the headland comprise slates of the Trevose Slate Formation, and intruded dolerite forms Park Head. (GS 444) (Photographer: A J J Goode)

BRITISH GEOLOGICAL SURVEY

E B SELWOOD
J M THOMAS
B J WILLIAMS
R E CLAYTON
B DURNING
O SMITH
L N WARR

Geology of country around Trevose Head and Camelford

CONTRIBUTORS

Trevose Head and St Breock Downs area geology
A J J Goode
B E Leveridge

Basic igneous rocks
P A Floyd

Bodmin Moor Granite
C S Exley

Economic geology
C M Bristow
R C Scrivener

Hydrogeology
M A Lewis

Research commissioned by the
NATURAL ENVIRONMENT RESEARCH
COUNCIL from the UNIVERSITY OF
EXETER

Memoir for 1:50 000 Geological Sheets 335 and 336 (England and Wales)

This memoir, and the 1:50 000 map that it describes, are the product of a contract between the Natural Environment Research Council and the University of Exeter. The interpretations presented are those of the authors.

London: The Stationery Office   1998

ISBN 0 11 884514 4

Bibliographical reference

SELWOOD, E B, THOMAS, J M, WILLIAMS, B J, CLAYTON, R E, DURNING, B, SMITH, O, and WARR, L N. 1998. Geology of the country around Trevose Head and Camelford. *Memoir of the British Geological Survey*, Sheets 335 and 336 (England and Wales).

Authors

E B Selwood, BSc, PhD
J M Thomas, BSc, PhD
B J Williams, BSc
R E Clayton, BSc, PhD
B Durning, BSc, PhD
O Smith, BSc
L N Warr, BSc, PhD
University of Exeter

Contributors

A J J Goode, BSc
B E Leveridge, BSc, PhD
British Geological Survey, Exeter

P A Floyd, BSc, PhD
University of Keele

C S Exley, MA, DPhil
University of Keele

C M Bristow, BSc, MSc
Camborne School of Mines

R C Scrivener, BSc, PhD
British Geological Survey, Exeter

M A Lewis, BA, MSc
British Geological Survey, Wallingford

Other publications of the Survey dealing with this district and others in the region

BOOKS
British regional geology
South-West England,
4th edition 1975 Reprinted 1985
Memoirs
Tintangel and Bude Coast (322) 1972
Boscastle and Holsworthy (322, 323) 1973
Tavistock and Launceston (337) 1911*
Newquay (346) 1906*
Bodmin and St Austell (347) 1909*
Plymouth (348) 1907*

Mineral Reconnaissance Programme reports
No. 103 Mineralisation north of Wadebridge, Cornwall
No. 108 Geochemical investigations near Camelford, Cornwall

MAPS
1:1 000 000
Geology of the United Kingdom, Ireland and adjacent continental shelf (South sheet)
1:584 000
Tectonic map of Great Britain and Northern Ireland
1:250 000
South Sheet (Geological)
South Sheet (Quaternary)
South Sheet (Aeromagnetic)
1:250 000
Lands End (Solid geology)
Lands End (Sea bed sediments)
1:50 000 and 1:63 360
Sheet 322 (Boscastle) (1964)
Sheet 323 (Holsworthy) (1974)
Sheet 337 (Tavistock) (1993)
Sheet 346 (Newquay) (1906)
Sheet 347 (Bodmin) (1909)
Sheet 348 (Plymouth) (1907)

* *out of print*

Printed in the United Kingdom for The Stationery Office
J29824   C6   1/98

CONTENTS

FIGURES

TABLES

PLATES

ACKNOWLEDGEMENTS

NOTES

The map which this memoir describes was largely surveyed by researchers at the University of Exeter as part of a Natural Environment Research Council mapping contract. The mapping was carried out by Dr R E Clayton, Dr B Durning, Miss O Smith and Dr L N Warr under the supervision of Dr E B Selwood and Dr J M Thomas. The remaining area was surveyed by Mr A J J Goode and Dr B E Leveridge of the British Geological Survey (BGS) under the direction of Dr R W Gallois, Regional Geologist. Drs M T Holder and B E Leveridge acted as the BGS/University liaison officers for the project.

The memoir has been largely written by Drs Selwood and Thomas with the assistance of Mr B J Williams, using the Technical Reports and PhD theses prepared by the university surveyors. Contributions have also been included from Mr Goode and Dr Leveridge describing the area that they surveyed; from Dr P A Floyd on the basic igneous rocks and associated metamorphism, Dr C S Exley on the granite and the associated minor acid intrusive rocks, Professor C M Bristow on the china-clay deposits, Dr R C Scrivener on the metalliferous mineralisation and Mrs M A Lewis on the hydrogeology.

In addition to the named contributors, expert advice was given by Professor M R House on macrofaunas, Dr B Owens and Dr A Dean on palynomorphs, Dr B Rice-Birchall on basic igneous rocks, and Dr R W O'B Knox on the sedimentology of the Trevose Slate Formation. The photographs were taken by Mr G F Logan.

The authors gratefully acknowledge information and assistance given by South West Water Plc for hydrogeological data; Cornwall County Council for site investigation data; English China Clay International Ltd, Delabole Slate Ltd, Natural Stone Products Ltd (Hantergantick Quarry), and Dimensional Stone Ltd (De Lank Granite Quarries) for details of the extractive industry. In addition help given by numerous farmers and landowners throughout the district during the course of the geological survey, is gratefully acknowledged.

The word 'district' used in this memoir means the area included in 1:50 000 scale Geological Sheet 335 and 336 (Trevose Head and Camelford).

Figures in square brackets are National Grid references; places within the Trevose Head and Camelford district lie within the 100 km squares SW and SX. The grid letters precede the grid numbers.

The authorship of fossil species is given in the index of fossils.

Numbers preceded by A refer to photographs in the Geological Survey collections.

Numbers preceeded by the letter E refer to thin-sections in the collection of the British Geological Survey.

PREFACE

The Trevose Head and Camelford district of north Cornwall is a rural area of outstanding natural beauty, encompassing a range of landscapes which reflect the variety of the underlying rocks and the erosive processes that have acted upon them. The uplands of the Bodmin Moor Granite rise sharply from the surrounding country rocks in the eastern part of the district, to dominate the inland scene. The treeless tableland flanking the moor, conceals deeply incised, wooded valleys of great charm which are rarely discovered by visitors. Such valleys were formed when sea-level fell during periods of glacial advance in northern Europe. Now that the climate has ameliorated, the sea-level has risen to drown the lower reaches of the river system, and the sea has been carried deep into the countryside through the Camel Estuary. This ria is rapidly silting up to give mud flats and sand banks rich in wild life.

This, the second edition of a memoir originally published in 1910, has been completely rewritten and incorporates accounts of all the geological formations in the district, the granite, and the resources of the extractive industry. It is a product of a Natural Environment Research Council Mapping Contract with the University of Exeter between 1986 and 1991. The memoir embodies the results of mapping and research by staff and PhD students of the Exeter Geology Department, and includes the input of specialists from Keele University and Camborne School of Mines. It is pleasing that the BGS has played a key collaborative role throughout, firstly in providing the support of our own specialist expertise and subsequently in overseeing production of the map and memoir through to successful completion.

The district records the development of the Trevone Basin in which some 6000 m of Devonian marine sediments accumulated. This basin represents an area of persistent downwarping, developed on a broad continental shelf extending southwards from the Caledonian Continent. Such subsidence was achieved by continued movements on deep-seated fractures in the Earth's crust. These faults also provided pathways allowing great quantities of magma to reach the surface and form pillowed lava flows and tuffs.

After about 70 million years of almost continuous sedimentation, the contents of the basin were squeezed out as a deformation front spread northwards across the region, late in Carboniferous times. Folding and thrusting with opposed directions of transport developed, to give a geological structure of great complexity. The last major event associated with this mountain building episode was the intrusion of the Bodmin Moor Granite about 290 million years ago. As the magma cooled, volatile-rich liquids deposited metallic ores within and adjacent to the slowly cooling granite. At about the same time, hot water circulating through the rock led to the local alteration of the granite and the formation of kaolin-rich china clay. Few younger deposits have survived in the district. Planation surfaces record repeated marine transgressions in the Pliocene and Pleistocene which reduced a more rugged topography to the present landscape.

The tranquillity of the landscape belies a dynamic and complex earth history extending back some 400 million years. This memoir enables us to better understand that history.

Peter J Cook CBE, DSc, CGeol, FGS
Director

British Geological Survey
Keyworth
Nottingham
NG12 5GG

ONE

Introduction

GEOGRAPHICAL SETTING

The Trevose Head and Camelford district has varied and attractive scenery; it includes the rugged, wave-torn cliffs that face into Atlantic storms, and the high, rolling, tor-studded uplands of Bodmin Moor. Both contrast with the steep-sided, sheltered, wooded valleys of the rivers Allen, Camel and Fowey (Figure 1). In the south-east, the rolling uplands rise to 200 m above OD on St Breock Downs, and fall gently westwards towards the coast where the Staddon Grit forms impressive cliffs with small sandy coves. Alternating cliffs and bays continue northward past the lighthouse on Trevose Head to Stepper Point. There, a dolerite body has resisted wave erosion, and the headland protects the mouth of the Camel estuary and the coast at Polzeath from south-westerly storms. The offshore islands, from The Quies off Dinas Head to The Mouls off The Rumps, are made of similar resistant rocks; they add much interest to the coastal scenery and provide safe resting places for many sea birds.

Inland, the area south of the Camel estuary comprises undulating farmland cut by steep valleys. The fishing town of Padstow and the market town of Wadebridge, at the lowest bridging point of the river, both nestle in sheltered positions on the south bank of the estuary. Similar gently sloping farmland rises from the north bank of the estuary, but to a lower altitude than that south of the estuary. Large quantities of sand have been blown inland from the beaches of the estuary; the Church of St Enodoc, near Rock, burial place of Sir John Betjeman, was formerly covered by blown sand. Between the Camel estuary and the northern boundary of the district, resistant volcanic rocks give rise to areas of higher ground, and to the dramatic high cliffs at Pentire and Rumps points, Kellan and Varley heads. Softer slates commonly back sandy bays, such as Hayle Bay at New Polzeath and Lundy Bay on the north-facing coast east of Rumps Point. Lacking the harder igneous rocks of the Pentire area, coastal erosion has proceeded much farther east in Port Isaac Bay, where slates form the cliffs from Bounds Cliff to Trebarwith Strand. Rejuvenated streams have cut their valleys deeply into the high-sea-level platforms of the cliff-top plateau. Weak zones in the rocks have been exploited by coastal erosion to give narrow inlets, within which the small harbours of Portquin, Port Isaac and Portgaverne are situated, their houses cramped into narrow, steep-sided valleys.

The rather austere and exposed plateau fringing the coast, falls towards the south-east, where the wooded valleys of the southward flowing rivers Allen and Camel have been deeply incised. In the more sheltered landscape mixed farming flourishes. East of the Camel valley, which closely follows the margin of the granite (Figure 2), Bodmin Moor rises abruptly and the scenery changes to rectilinear patterns of large-scale, rounded ridges and flat-bottomed peaty valleys. Scattered tors and broad benches of granite crop out, and large boulder-fields characterise the landscape. Much of the moor has been enclosed and extensive areas have been agriculturally improved to give much coarse grassland vegetation. The high altitude and rainfall have allowed the development of thick peat in the valley floors, but only thin upland peat over the rolling hills.

The coastal plateau extends to the northern margin of the district and sweeps around the north-western limit of Bodmin Moor to give the flat landscape of Davidstow Moor. Farther east, there is a gradual slope towards the headwaters of the River Tamar, where the shelter offered by the high moorland lying to windward ameliorates the climate in the incised valleys. The southern margin of Bodmin Moor presents a much more enclosed, sheltered pattern of heavily wooded, deeply inset valleys. Many small villages and hamlets are tucked away in these sheltered locations; more open, exposed farmland and a few somewhat improved commons occupy the intervening hill-tops.

Through agriculture, man's effect on the landscape is ubiquitous. Even Bodmin Moor is split up by granite stone walls, and grassland has been improved so that little natural heathland remains. Numerous Neolithic hut circles testify to the long history of human habitation on these uplands.

The large, white, waste heaps of china clay from the Stannon Pit in the north and Park Pit in the south of Bodmin Moor are visible over long distances. Smaller, abandoned china clay pits, now flooded, are scattered over the moor with their waste heaps overgrown. The effects of ore extraction from the metalliferous mines create a less obvious impact on the modern landscape. Early alluvial tin working affected almost all the valleys and depressions, and the opencast working of fractured, subsurface ore-bearing rock is marked by long deep trenches, now much overgrown and degraded, cutting across the hillsides. Granite has been extracted in large quantities from the moor, largely for building material; first by working the boulder fields and later by solid rock quarrying. Although most workings are now abandoned, two modern quarries near the western margin still extract granite for building and decorative stone work. In the past, vertical dykes of elvan were preferentially quarried; where fresh, the close-spaced jointing was exploited for breaking as road metal, but where deeply weathered, the rock could be sawn and used as ashlar stone.

Much slate has been exploited for roofing and walling use, particularly at coastal localities. Most quarries are now abandoned, but weather-stained 'rustic' slate is still

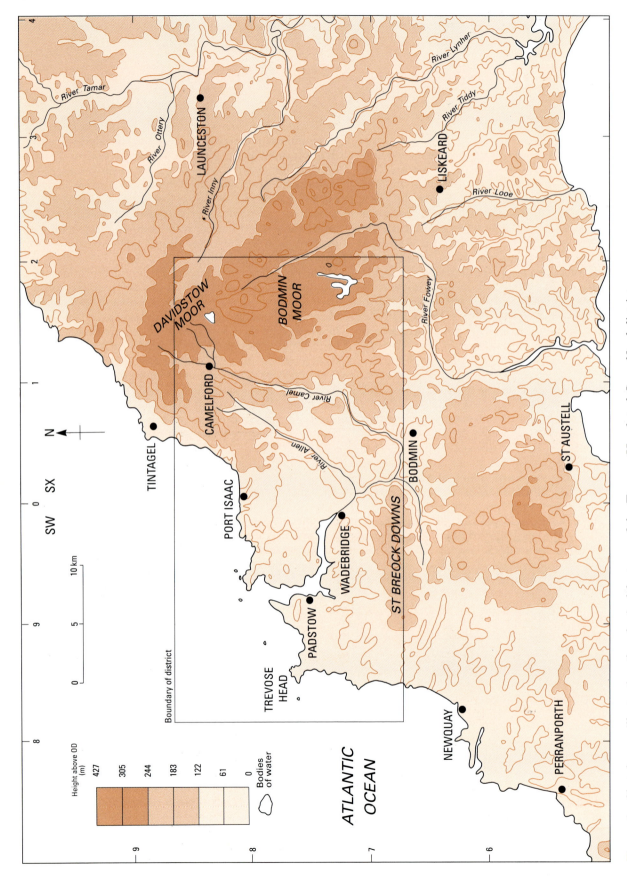

Figure 1   Sketch map illustrating the physical features of the Trevose Head and Camelford district.

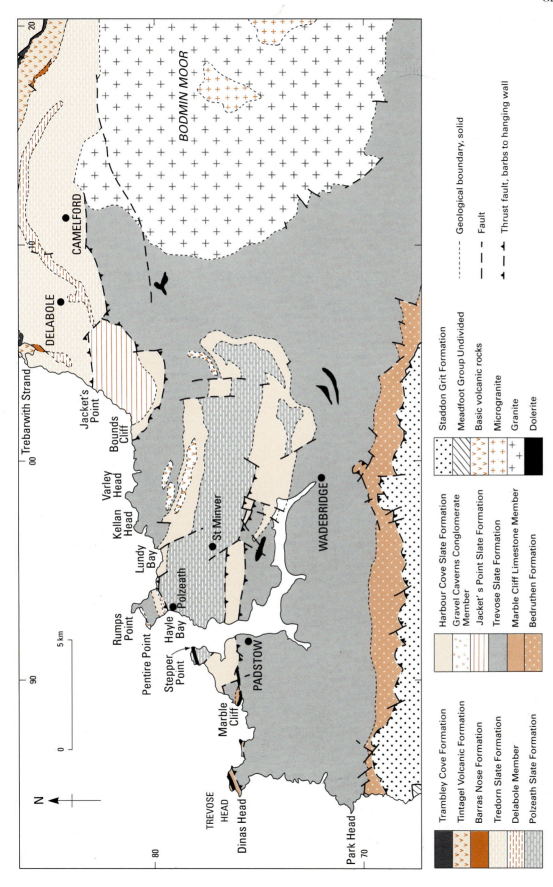

Figure 2   Sketch map of the solid geology of the district.

Trambley Cove Formation
Tintagel Volcanic Formation
Barras Nose Formation
Tredorn Slate Formation
Delabole Member
Polzeath Slate Formation

Harbour Cove Slate Formation
Gravel Caverns Conglomerate Member
Jacket's Point Slate Formation
Trevose Slate Formation
Marble Cliff Limestone Member
Bedruthen Formation

Staddon Grit Formation
Meadfoot Group Undivided
Basic volcanic rocks
Microgranite
Granite
Dolerite

Geological boundary, solid
Fault
Thrust fault, barbs to hanging wall

worked near Trebarwith Strand, and high-grade roofing slates are obtained from the famous Pengelly Quarry at Delabole. Waste slate from the extensive tips at Delabole is used as general fill material throughout the area.

Height and high rainfall give Bodmin Moor useful potential as a water resource and storage area. The reservoirs at Crowdy and Colliford have enhanced the local scenery and now offer varied amenities. Dozmary Pool, one of the few natural water bodies on Bodmin Moor, and Bolventor (Jamaica Inn) are popular tourist sites, largely due to their literary connections.

Fishing and some coastal shipping still operate out of Padstow Harbour, but the smaller harbours elsewhere are geared to tourism. Holiday-related development has taken place in many places along the coast south from Pentire Point, where the beaches and cliff scenery are particularly popular. Wadebridge and Camelford have developed as the main service-industry centres for the region, but face competition from Bodmin, lying just south of the district, which has better transport connections.

PREVIOUS RESEARCH

The Trevose Head and Camelford district was included in the first regional survey in Britain conducted by Sir Henry De la Beche (1839). The Bodmin Moor granite was differentiated, and all country rocks were referred to the 'Grauwacke Group'. Within this group 'Trappean rocks' and elvans were distinguished. Areas of blown sand, and the disposition of rocks around the structure which has come to be known as the Davidstow Anticline were also noted. In this district, which was not involved in the controversy surrounding the creation of the Devonian system (see Rudwick, 1985), the strata of the 'Grauwacke Group' were accepted into the Devonian without question; a judgement that was largely vindicated by the later discovery of diverse invertebrate faunas. Among early collectors Howard Fox was pre-eminent. His and other early palaeontological work was reviewed and supplemented by a wealth of new data in the first edition of this memoir (Reid et al., 1910), a volume which has long provided the only comprehensive description of the inland areas and which remains a valuable work of bibliographical reference and local detail. The geological maps (One-Inch Sheets 335 and 336 published in 1907) that accompany the memoir delimited Lower, Middle and Upper Devonian strata across the district for the first time. In the absence of evidence to the contrary, lithostratigraphical and chronostratigraphical boundaries were shown to be coincident. Only the Staddon Grits and Meadfoot Beds were separated as named lithostratigraphical units.

Within this sequence, volcanic rocks and basic intrusive rocks feature prominently, and prompted some of the earlier descriptions of pillow lavas in Britain (Whitley, 1849; Reid and Dewey, 1908; Dewey and Flett, 1911). The spectacular development of spilosites and adinoles associated with many of the basic intrusions in the district featured in early works (Fox, 1895b; Reid et al., 1910; Dewey, 1915), and in the classic studies of Agrell (1939, 1941).

Geochemical studies by later workers have shown that the basic volcanic rocks of north Cornwall constitute a distinct magmatic province of predominantly alkali basalts (Floyd, 1982; Floyd et al., 1983; Floyd, 1984; Rice-Birchall and Floyd, 1988) which feature characteristics of within-plate basalts (Floyd, 1982; Rice-Birchall, 1991).

Investigations into conditions of low-grade metamorphism have exploited the sensitivity of certain mineral species to the physical changes associated with regional deformation (Primmer, 1985a; Robinson and Read, 1981). The mapped distribution of volcanic rocks gave support to the concept (Reid et al., 1910) of a regional syncline controlling the distribution of strata throughout the district. This structure was confirmed by House (1956) when he recognised the repetition of Middle Devonian strata across the fold. This seminal work led to a progressive reappraisal of the stratigraphy and palaeontology of the area (House, 1961, 1963; Gauss and House, 1972; Beese, 1982, 1984; Selwood and Thomas, 1986a; Austin, et al., 1992) and to the recognition of most of the formations described in the present memoir.

Mapping by the Geological Survey on Sheet 332 (Boscastle), adjacent to the present district, distinguished the Tintagel Succession (Freshney et al., 1972, McKeown et al., 1973). The delimitation of these formations on the northern part of the Camelford Sheet by Mr G Bisson and Dr M C McKeown in 1964, revealed Carboniferous strata within the Camelford district for the first time. These revisions, and those indicated by House, were incorporated into new one-inch editions of Sheet 336 in 1969 and Sheet 335 in 1970.

There have been notable investigations of biostratigraphically important groups in the district; goniatites by House (1956, 1963), ostracods by Gooday (1973), and conodonts by Kirchgasser (1970), Mouravieff (1977) and House et al. (1978). All of these studies have furthered the stratigraphical understanding of the region. The Marble Cliff succession has received international recognition because it reveals one of the more complete Middle to Upper Devonian conodont sequences known.

There has been much discussion about the large-scale structure of the district. The concept of a regional synclinal structure was reinforced by analysis of small-scale structures in coastal sections that cut across the regional strike. Gauss (1966, 1967) identified polyphase deformation and an area of opposed fold-facing lying north of Padstow (the 'Padstow Confrontation' of later authors), south-facing to the north and north-facing to the south. He suggested that the variation in facing was due to tectonic transport by gravity folding, during which two blocks were brought into juxtaposition by movement along a thrust (Gauss, 1973 p.308). Such tectonic shortening was consistent with the contiguity of contrasting, coeval successions recognised by Gauss and House (1972) in the northern (Pentire Succession) and southern (Trevone Succession) limbs of the regional syncline.

Roberts and Sanderson (1971, p.87) concluded that structures south of the district formed at an earlier date than those to the north, and that the Padstow Confrontation resulted from two phases of deformation. The early phase of folding was envisaged to die out northwards in the region of Polzeath; these early folds were refolded by a southwards-transporting deformation. Roberts and Sanderson (1971) thus recognised a zone of confrontation, as opposed to the surface of confrontation identified by Gauss (1973).

Later interpretations of the structure of the district can be viewed as refinements of the models proposed by either Gauss or Roberts and Sanderson. It is now generally agreed that the regional syncline, the St Minver Synclinorium, is the product of multiphase deformation.

A regional overfold model was developed for the north Cornwall coast (Hobson and Sanderson, 1975; Sanderson, 1979; Rattey and Sanderson, 1982; Ferguson and Lloyd, 1982; Hobson and Sanderson, 1983; Andrews et al., 1988) which proposed the existence of a major south-closing fold nappe, incorporating a conformable Middle Devonian to Upper Carboniferous succession between Padstow and Bude. The transition from upright folding at high structural levels, to recumbent folding at lower levels was held to reflect the suprastructure and more highly strained infrastructure of a major south-facing fold nappe (the Millook Nappe of Rattey and Sanderson, 1982). This infrastructure was noted to extend as far south as Padstow; there, D_1 south-facing structures were indicated overprinting D_1 north-facing structures within the Padstow Confrontation Zone.

Shackleton et al. (1982) interpreted structures across south-west England using a thin-skinned model, with a basal decollement dipping gently south. Within the district, the south-facing structures were suggested to have been generated in a zone of reversed shear (back-thrusting) in which north- and south-facing zones were separated by a north-dipping thrust.

Selwood et al. (1985) and Selwood and Thomas (1986a) noted that the overfold model was not compatible with the stratigraphy. They recognised a number of coeval successions juxtaposed by thrusting, and proposed a southerly derived thrust-nappe model for the north Cornwall coast, similar to that proposed for structures in the Tavistock (Sheet 337, Isaac et al., 1982) and Newton Abbot (Sheet 339, Selwood et al., 1984) districts.

More recent structural analyses (Andrews et al., 1988; Pamplin and Andrews, 1988; Durning 1989a and b; Warr, 1988) have reaffirmed that the D_1 structures between the Rusey Fault Zone and the Padstow area are south-facing. These authors have ascribed the D_1 structures to a regional backfolding and thrusting event. Regional northward transport subsequently took place in early D_2 (ductile) and late D_2 (brittle) thrusting. Andrews et al. (1988) and Pamplin and Andrews (1988) believed that D_1-south predated D_1-north, and identified a transitional confrontation zone. In contrast, Durning (1989b) and Warr and Durning (1990) thought that D_1-north predated D_1-south, and that an abrupt change of facing takes place across a major thrust.

Compared with the other plutons of the Cornubian batholith, the Bodmin Moor granite is remarkably uniform; the greater part of the outcrop is made up of coarse-grained biotite granite. Reid et al. (1910) mapped several areas of fine-grained granite, and Ghosh (1927) argued for the occurrence of multiple intrusive phases of coarse granite types. It has also been shown (Hawkes and Dangerfield, 1978; Dangerfield and Hawkes, 1981) that a subdivision of granite varieties can be effected on the abundance of megacrysts.

In a geochemical study, Edmondson (1972) used trend-surface analysis to demonstrate that the concentration of a number of the major and minor elements in the granite show a roughly concentric distribution. The chemical, normative and modal compositions reveal the body to be a characteristic S-type granite (Chappell and White, 1974), derived by the deep crustal melting of sedimentary rocks. Rb/Sr dating (Darbyshire and Shepherd, 1985) at 287 ± 2Ma, established the granite as one of the older plutons in the batholith.

The importance of deep-seated, north-west-trending, vertical faults that cut the cupola was noted by Dearman (1963); Exley (1965) commented that these are paralled by vertical joint sets. Reid et al. (1910) identified an important fault traversing the north-western part of the moor between Devil's Jump [SX 1009 8016] and Lanlavery Rock [SX 1562 8261]. Significant downthrow to the north-west is believed to have preserved granite close to the roof of the pluton.

Within the Quaternary sequence of the district, strand levels have been documented by Clarke (1963) and Everard et al. (1964). More attention has focused on the superficial deposits of the Camel Estuary, particularly the boulder gravels (Arkell, 1943; Clarke, 1973; Source, 1985), which have been variously ascribed to a glacial or glacial-outwash origin.

GEOLOGICAL HISTORY

Devonian sedimentation was initiated on continental crust forming the southern extension of the Caledonian continent of Eastern Avalonia. At first, thick continental sediments extended southwards across the region towards an ocean lying farther south. Such sequences are developed within the Dartmouth Slates cropping out immediately south of the district. The Meadfoot Group, the oldest rocks represented in the district, records the initiation in Emsian times, of a progressive onlap of marine, shallow-water, clastic sediments. The Staddon Grit Formation, which is extensively developed at the top of the group, characteristically shows evidence of storm-induced processes. Such activity continued during the deposition of the overlying Bedruthan Formation, which records an upwards decrease in sandstone content.

The Trevone Basin is believed to have developed during the Devonian as an extensional half-graben, bounded to the north and east by deep-seated basement structures which appear to have controlled basic magmatism during sedimentation.

Differential subsidence of the shelf in mid-Devonian times led to the accumulation of a thick basinal sequence, the argillites of the Trevose Slate Formation. A barred basin with periodic bottom water anoxia is indicated; the Marble Cliff Limestone Member includes fine-grained, carbonate turbidites probably derived from a sublittoral, intrabasinal bar lying south-west of the district. Basin subsidence was accompanied by intrusive and extrusive volcanic activity which was particularly intense at the northern margin of the basin. Deep-water shelf argillites, of the Jacket's Point Slate Formation, characterised the inner (northern) shelf at this time.

The northward spread of open, progressively deeper-water basinal conditions, and a marked reduction in volcanic activity across the area is recorded by the Harbour Cove Slate and Polzeath Slate formations.

Basin inversion occurred during a north–south compressional event in the late Visean/early Namurian, and was probably controlled by pre-existing extensional structures in the basement. In the northern part of the Trevone Basin, the geometry of the inversion, involving large-scale, downward and upward, south-facing structures, is believed to represent a back-thrusted and back-folded wedge initiated by buttressing against fault-bounded blocks developed at the northern margin of the basin. It is not possible to determine if the entire contents of the northern parts of the basin were back-thrusted, or whether more localised backthrusting was associated with the reactivation of antithetic faults, originally developed during basin extension. The southern parts of the basin underwent simple northwards verging folding and thrusting.

Following the D_1 deformation, a phase of out of sequence D_2 thrusting towards the north-north-west was initiated across the inverted basin. These thrusts dismembered the F_1 regional folds into slices, which now largely control the disposition of stratigraphic units. Local lateral ramp geometries, and oblique thrust traces are evident (Andrews et al., 1988, Warr, 1988). Localised non-coaxial overprinting of deformations within the footwall of D_2 thrusts, led to complex zones of refolding and multiple cleavage development, and this might be the origin of the Padstow Confrontation (Warr, 1993).

The Tintagel Unit forms the highest structural unit, and is carried on a complex of thrusts and late north-dipping extensional faults. It is variously argued that the Tintagel Unit was derived initially from the north (Warr, 1989) or south (Selwood and Thomas, 1986a; Selwood, 1990).

The intrusion of the Bodmin Moor granite postdated the main Variscan orogenic events and has been radiometrically dated at 287 ± 2Ma (Darbyshire and Shepherd, 1985). Associated acid porphyry intrusions (elvans) have yielded dates of 282 ± 6Ma and 270 ± 9Ma (Darbyshire and Shepherd, 1985).

Metalliferous mineralisation in the district shows a long and complex history, starting with stratiform sulphides within the black slates of the Staddon Grit, Trevose Slate and Barras Nose formations. Two main episodes of vein mineralisation are reported; a tectonically concentrated antimony-arsenic-gold assemblage, which is particularly associated with the volcanic sequences in the northern part of the Trevone Basin, and the tin-wolfram-copper hydrothermal mineralisation, which is linked to granite emplacement. Tourmalisation, greisenisation and kaolinisation of the granite form part of a continuum of hydrogeochemical activity which started in the late-magmatic stage of granite crystallisation and has been maintained until the present (Bristow et al., in press).

Apart from granite regolith formation, which may well have been initiated in Tertiary times, superficial deposits constitute the only record of post-Variscan geological history.

TWO

Devonian and Carboniferous

The stratigraphy of the district is magnificently displayed in coastal sections which cut across the strike of the beds. These strata have been the subject of detailed analysis, but poor exposure has left the inland geology relatively unexplored. The stratigraphy identified by earlier workers on the coast cannot be maintained inland, and in the present survey a number of stratigraphical units previously described as formations have therefore been reduced to member status within more broadly defined formations.

Although the sequence across the sheet was originally conceived as conformable, though highly faulted (Reid et al., 1910), later workers recognised the presence of distinct, coeval successions. Four successions* were recognised during the resurvey of the district; the Padstow Succession (the Trevone and Pentire successions of Gauss and House, 1972), the Tintagel Succession (after Freshney et al., 1972), the Bounds Cliff Succession (after Selwood and Thomas, 1986a; Warr, 1993), and the Liskeard Succession (after Burton and Tanner, 1986, (Figures 3 and 4). The primary relationship between these successions has in part been obscured by later tectonism, but a stratigraphical model can be proposed which is consistent with structural data. Current biozone schemes for the Devonian are shown in Figure 5.

PADSTOW SUCCESSION

The Trevone and Pentire successions were erected by Gauss and House (1972) to describe the variations in facies in the Devonian sediments recorded along the coastal section, traversing both southern and northern limbs of the St Minver Synclinorium (Figure 8). This difference was lost in the latest Devonian (Famennian) when mud, now forming the purple and green Polzeath Slates, was deposited across the whole region. Gauss (1973) and Durning (1989a) have suggested from sedimentological evidence that those two successions were deposited in different areas, and were brought together tectonically. The Pentire Succession contains evidence for much tectonic instability during its accumulation, and is distinguished by a thick succession of volcanic rocks, associated with much soft-sediment deformation. In contrast the Trevone Succession was described by Gauss and House (1972) as fully basinal; with mudrock sequences including carbonate and distal arenaceous turbidites, lavas and volcanoclastic rocks.

During the present survey, lithostratigraphical variations around the St Minver Synclinorium were accommodated in terms of the Trevone Succession alone. The Trevose Slate Formation has been expanded to include the Pentire Pillow Lava Group, and the Gravel Caverns Conglomerate and Pentire Slates are included in the Harbour Cove Slate Formation. This last has been mapped on the northern limb of the St Minver Synclinorium for the first time. The Marble Cliff Beds have been given member status, but the Merope Island Beds and Longcarrow Cove Beds have not been differentiated. The revised unified succession is here termed the Padstow Succession. The correlation of the Givetian to Famennian formations depicted within the district are given in Figure 3. The original concept of the Trevone Succession has been further modified so that the Padstow Succession includes the Lower Devonian formations which underlie the Trevose Slate Formation.

Meadfoot Group

In the type area in south Devon, the Meadfoot Group comprises a sequence of marine, shallow-water clastic sediments, constituting a mainly argillaceous Bovisand Formation, overlain by the Staddon Grit Formation (Seago and Chapman, 1988). The latter has been interpreted as a southward prograding deltaic sequence (Pound, 1983). These formations broadly correspond to the Meadfoot Beds and Staddon Grits represented on the first edition 1:63 360 Geological Sheets of south Devon and Cornwall.

The Staddon Grits were mapped (Reid et al., 1910) through the high ground at the southern margin of the present district, but the Meadfoot Beds were restricted to an area of less than 1 km² in the south-west of the area near Mawgan Porth.

Rocks of the Meadfoot Group crop out near Mawgan Porth [SW 848 676] where they are mainly medium to dark grey slaty mudstones with sporadic interbedded thin graded siltstone and pale green silty tuff beds 0.1 to 0.2 m thick. X-ray diffraction analyses of the tuffs show the presence of abundant plagioclase suggesting a basic to intermediate composition of the source magma. Graded bedding in some siltstone units indicates that in part at least, this sequence is upside down. Because important facies changes are believed to occur along strike, these beds have not been assigned to a formation in the present survey. The fauna was reviewed by House and Selwood (1966). An Emsian age for the group has been deduced from brachiopod faunas (Evans, 1981; 1983; 1985).

Staddon Grit Formation

The Staddon Grit crops out along the coast for about 1.25 km between a steeply faulted boundary with undif-

* The term 'Succession' is used in this memoir to denote a rock sequence that differs in facies from, but is equivalent to or overlaps in age with, other sequences in the district. All the successions are juxtaposed tectonically and their original relative dispositions are unknown in detail. Usage of the term follows Gauss and House (1972).

Figure 3 Stratigraphical successions in the Devonian and Carboniferous rocks of the district; GCCM = Gravel Caverns Conglomerate Member. MCM = Marble Cliff Limestone Member. * Not exposed in district. θ Transition Group of Freshney et al. (1972) renamed in Selwood and Thomas (1993). † After Burton and Tanner (1986).

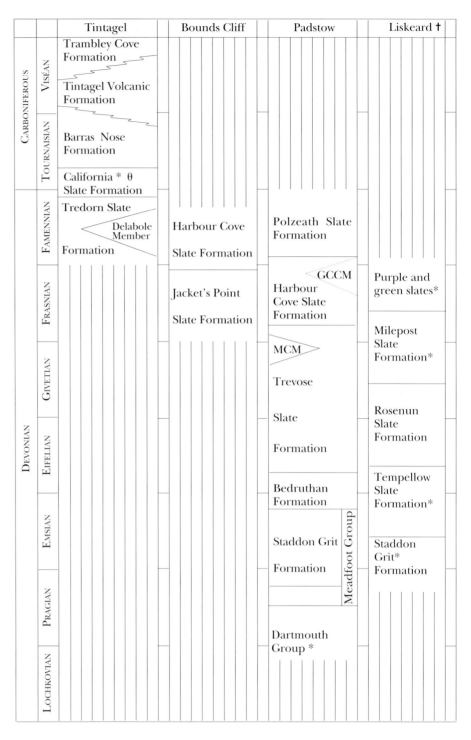

ferentiated Meadfoot Group sediments at Trenance [SW 8492 6768], and a gently inclined south-dipping fault at Whitestone Cove [SW 8471 6900] and Carnewas Point [SW 8473 6896] where it overlies the Bedruthan Formation. The stratigraphical thickness of the Staddon Grit cannot be determined accurately in the district, but it is estimated to be 400 m in the coastal section.

Inland, the Staddon Grit strikes eastwards forming a narrow tract of high ground including Bear's Downs [SW 897 680], Trelow Downs [SW 920 685], Scotland Corner [SW 945 680], St Breock Down [SW 970 685] and Great Grogley Downs [SX 010 678]. The northern boundary of the formation dips at a low angle to the south and is inferred to be a thrust. In the east of the outcrop, sandstone beds range in thickness between 60 mm and 0.6 m, and show varied geometries; parallel-sided planar beds, non-planar bounded beds and beds with basal scours are all present.

The formation comprises fine- to medium-grained sandstones interbedded with laminated siltstone and mudstone beds. On the coast, sandstones may approach 3 m in thickness, but average 0.3 m. The individual beds commonly have sharp erosional bases and slightly gradational tops. They are laminated, and locally display wavy bedding and asymmetrical ripples.

The sandstones are invariably interbedded with thin (less than 10 mm), fine-grained, green-buff sandstones and mudrocks. These thinly bedded lithologies have undergone minor recrystallisation during deformation and metamorphism, which has modified the primary sedimentary structures.

Discontinuous sandstone lenses, cross-lamination and parallel-lamination sets are evident in the coarser facies, with mudrock laminae forming thin intervening units between impure lithic arenite and siltstone. Greenish grey slates occur in the lower levels of the formation exposed in the district; they carry a variable siltstone content and a few sandstone laminae.

Sandstone component grains include quartz, altered indeterminable feldspar, white mica, opaque minerals and lithic grains. The lithic component includes polycrystalline and strained quartz, quartzite and lithic tuff. The proportions of the components indicate compositions varying between lithic arenite and sublithic arenite.

Sporadic bioturbation structures indicate periodic colonisation by burrowing organisms during periods of oxic bottom waters. The presence of the palynomorphs (Plate 1) *Dictyotriletes emsiensis* (Allen) McGregor, *Leiotriletes* sp. and *Retusotriletes* cf. *triangulatus* (Streel) Streel (Dr A Dean, personal communication) from Dunmere [SX 0489 6775], indicates an Emsian age for the Staddon Grit.

The fine, wavy bedding and basal scouring characteristic of much of the formation, suggest storm-induced processes, and the ripples suggest wave action. A shallow–marine depositional environment is indicated. At Plymouth, Humphries and Smith (1989), have recognised coarsening- and thickening-upward sequences in the Staddon Grit which have been attributed to a fluvially dominated, low-wave-energy, deltaic environment subject to local reworking into a series of offshore bars (Pound, 1983).

Bedruthan Formation

The Bedruthan Formation (Figure 2), as defined here, crops out in its type section along the coast for nearly 1 km, between a faulted boundary with the Staddon Grit Formation at Carnewas Point [SW 8473 6896], and its conformable junction with the overlying Trevose Slate Formation at an unnamed point [SW 8475 6988] north of Queen Bess Rock. This sequence embraces the greater part of the Bedruthan Slates recognised by Ripley (1964), from the Staddon Grit Formation at Carnewas Point to just north of Pentire Steps [SW 848 705]. Previously, this succession had been included within the Middle Devonian grey slates of Reid et al. (1910), and subsequently within the Trevose Slate of Kirchgasser (1970). Much of the coastal section in the Bedruthan Formation has a southerly overturned dip. The estimated thickness at the coast is 350 m; inland, the formation thickens to about 750 m in the eastern part of the district. The southern boundary of the formation is marked by the overthrust massive sandstones of the Staddon Grit Formation; but the northern boundary is gradational into the Trevose Slate Formation.

The Bedruthan Formation comprises grey-buff, fine-grained sandstone interlaminated and discontinuously interbedded with siltstone and grey-green mudstone. Normally graded beds, cross-lamination, parallel lamination, wispy bedding, sporadic sandstone lenses and load structures are evident, but these are almost all lost where tight folding has caused the amalgamation of sandstone units. Subordinate dark grey units of mudstone and siltstone also occur at several stratigraphical levels within the formation, with rare interbedded parallel-laminated sandstones.

The sandstone proportion tends to decrease upwards through the formation. Bioturbation is common and thin layers of bioclastic debris are sporadically developed. The wispy bedded sandstone and layers of bioclastic debris are

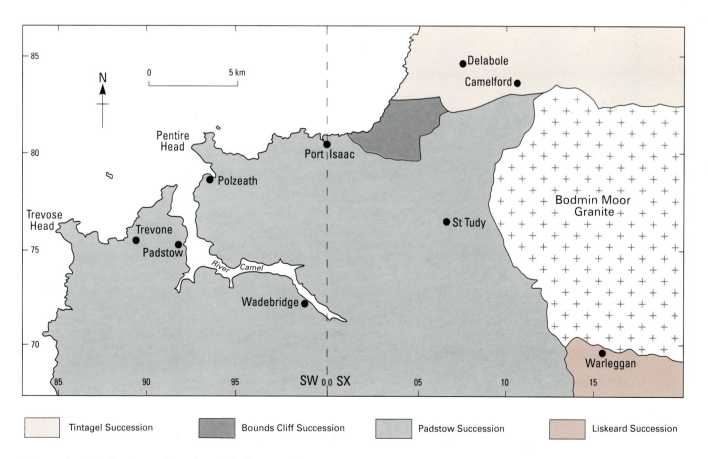

Figure 4 Distribution of stratigraphical successions.

Figure 5 Devonian conodant and ammonoid biozonal schemes: after Ziegler and Sandberg (1990), and Oliver and Chlupac (1991).

Ac = Acutimitoceras, An. = Ancyrognathus, Anc. = Ancyrodelloides, C. = Cymaclymenia, Cabr. = Cabrieroceras, Ch. = Cheiloceras, Cr. = Crickites, crist. = cristatus, Eogn. = Eognathodus, I. = Icriodus, K. = Kalloclymenia, Ko. = Keonenites, Maen. = Maenioceras, N. = Nowakia, O. = Ozarkodina, Or. = Ornatoclymenia, P. = Protoxclymenia, Pa. = Palmatolepis, Ped. = Pedavis, Pet. = Petteroceras, Pg. = Progonioclymenia, Ph. = Phariceras, Pi. = Piriclymenia, Pl. = Platyclymenia, Pol. = Polygnathus, Pon. = Ponticeras, Pro. = Prolobites, Ps. = Pseudoclymenia, Sch. = Schmidtognathus, Si. = Siphonodella, Sp. = Sporadoceras, T.k. = Tortodus kockelianus, Teich. = Teicherticeras, W. = Wocklumeria.

Series	Stage	Conodont biozone	Ammonoid biozone	
LATE DEVONIAN	Famennian	Pa. praesulcata	Wocklum.	Ac. carinatum / C. euryomphala / W. sphaeroides / K. subarmata
		Pa. expansa	Clymenia.	Pi. piriformis / Or. ornata / Pg. acuticostata / P. serpentina
		Pa. postera	Plat.	Pl. annulata / Pro. delphinus / Ps. sandbergeri
		Pa. trachytera		
		Pa. marginifera		
		Pa. rhomboidea	Cheiloc.	Sp. pompeckji
		Pa. crepida		Ch. curvispina
		Pa. triangularis	Manticoc.	Cr. holzapfeli
	Frasnian	Pa. gigas		Manticoceras cordatum
		An. triangularis		
		Pol. asymmetricus	Pharcicer.	Ko. lamellosus / Pet. feisti / Pon. pernai / Ph. arenicum / Ph. lunulicosta / Ph. amplexum
MIDDLE DEVONIAN	Givetian	Pa. disparilis		
		Sch. hermanni, Pol cristatus		
		Pol. varcus		Maen. terebratum
				Maen. molarium
	Eifelian	Pol. xylus ensensis		Cabr. crispiforme
		T.k. kockelianus		Pinacites jugleri
		Pol. costatus costatus		
		Pol. costatus partitus		
EARLY DEVONIAN	Emsian	Pol. costatus patulus		Anarcestes
		Pol. serotinus		
		Pol. inversus		Teich. discordans
		Pol. gronbergi		Anetoceras
		Pol. dehiscens		
	Pragian	Pol. pireneae		
		Eogn. sulcatus kindlei		
		Eogn. sulcatus		
	Lochkovian	Ped. pesavis / Anc. delta / O. eurekaensis / I. w. woschmidti		

typical products of storm wave action. The Bedruthan Formation was probably formed under shallow-marine conditions. A revision of the extensive macrofauna obtained from the Bedruthan Formation (Reid et al., 1910) by Professor M R House has indicated an Eifelian age. Palynomorphs from the following localities (Dr A Dean, personal communication) [SX 0535 6835] *Emphanisporites rotatus* (McGregor) McGregor, *Dibolisporites albitiensis* McGregor and Camfield, *Retusotriletes* sp. and [SX 0524 6819] *Retusotriletes communis* Naumova suggest that the formation ranges down into the Emsian.

Trevose Slate Formation

The term Trevose Slate was introduced by Kirchgasser (1970) to describe the Middle Devonian grey slates exposed on the coast between Trevose and Constantine Bay. It included the grey slate of Gauss (1966), the light grey slate of House (1961) and the 'c² slates' depicted on the early editions of Geological Sheets 335 and 336 (1907). The formation was described by Gauss and House (1972), who designated the coastal section on the west side of Trevose Head, between Constantine Bay [SW 8570 7505] and Mackerel Cove [SW 8503 7613] as the type section. Later, in a detailed sedimentological study, Beese (1982) divided the Trevose Slate of the coastal outcrops into three formations, but these units have not been confirmed during this survey. In this work the Trevose Slate Formation terminology of Gauss and House (1972) has been retained, but the formation has been redefined to include the Longcarrow Cove Beds and Marble Cliff Beds of House (1961), which are not mapped as distinct formations. Beese (1982) defined the base of the formation as a conformable junction with the Bedruthan Formation, at the top of a 1 m-thick bed of volcanic rock exposed at Lower Butter Cove [SW 8437 7104] but the base of the formation as defined here is seen, to the south in the cliffs [SW 8487 6993] near Queen Bess Rock. The formation is Eifelian to early Frasnian in age.

The estimated thickness of the Trevose Slate (Figure 2) is about 3900 m; this allows for faulting and mesoscopic-scale folding but not for extension due to flattening perpendicular to cleavage. Much of the sequence shows regular lamination comprising alternating grey and darker grey laminae. The paler laminae consist of fine-grained siltstone, and the darker laminae consist of silty mudstone. Paler laminae generally show sharper bases and gradational tops, so that the sequence comprises two-laminae couplets. The couplets, which are commonly less than one millimetre thick, are generally laterally persistent. On the basis of laminae thickness there is an indication that successive laminae are grouped in lamina-bundles.

Beds of structureless siltstone, from 0.03 m to 0.10 m thick, are regularly interbedded with the laminated rocks. Bedding surfaces are planar, and lower boundaries are sharper than upper boundaries. In clean, wave-washed exposures, the beds are seen to comprise dark blue-grey central parts that are more resistant to erosion, and marginal parts weathering grey-buff. Thicker beds commonly display a two-fold division, the lower part being distinguished by the presence of impersistent laminae of fine cross-laminated sandstone.

At the base of the formation, the dark grey mudstones are essentially homogeneous, showing only well-spaced, pale grey laminae which are interrupted by sporadic lenticular bands of black slate 0.01 to 0.03 m thick. Sporadic beds of fine-grained sandstone up to 0.12 m thick show internal planar lamination and low-angle cross-lamination. In the upper part of the succession, most prominently in the 500 m below the Marble Cliff Limestone, there are sporadic laminae and thin laminated beds of limestone up to 0.10 m thick. This sequence yields a diverse and commonly pyritised fauna.

Some parts of the sequence are marked by the presence of deformed units generally less than a metre thick. These display synsedimentary fold and fault structures, including imbricate thrust stacks, indicating displacement through slumping. These structures are truncated by the planar beds of overlying laminated siltstones.

Disruption of primary layering can also be ascribed to burrowing; all stages of disruption are represented, culminating in the homogenisation of the sediment. In the lower half of the formation diffuse remnants of laminae are all that may remain of bedding through sequences of several metres of rock. In the upper half of the formation, there is greater bedding continuity, and some sections are devoid of evidence of bioturbation. Much of the laminated siltstone has a scatter of fine bioclastic debris and is partly bioturbated, but there are largely homogenised pale siltstone beds, up to 0.5 m thick, with scouring at their bases, which contain an abundance of bioclasts. Such beds commonly host larger burrows, and burrow sections up to 0.03 m by 0.20 m have been recorded, some penetrating undisturbed laminated sediment at their base.

Above the Marble Cliff Limestone Member, 300 m of mudstones at the top of the Trevose Slate Formation were named the Longcarrow Cove Beds by Gauss and House (1972). Beese (1984) gave further details of the lithologies, and raised the unit to formational status with a type section in the cliffs at Longcarrow Cove [SW 894 768].

This part of the formation is predominantly laminated mudstone but includes more homogeneous beds than those underlying the Marble Cliff Limestone. Thin beds of limestone, lava, tuff and volcaniclastic rocks, up to 1.5 m thick occur sporadically. Some of the tuffs and limestones near the base of the succession are graded and cross-laminated. The limestones are consistently thinner than those within the underlying Marble Cliff Limestone.

Pale buff siltstone beds are locally developed towards the top of the Trevose Slate east of the Camel Estuary. These are associated with a 1 m-thick debris flow deposit, consisting of, matrix-supported clasts of siltstone and shale up to to 0.03 m in diameter.

VOLCANIC ROCKS WITHIN THE TREVOSE SLATE FORMATION

There is a marked increase in the proportion of volcanic rocks in the upper part of the formation, with a variable distribution across the St Minver Syncline. Most occur on the northern limb of the structure, and east of the Allen

Valley Fault where lava flows, pillow lavas, hyaloclastites and breccias, such as those at Pentire Head, represent proximal volcanicity. Relatively few occur on the southern limb of the synclinorium. Of the latter, few are exposed in the coastal sections in the Trevose Slate, apart from those at Longcarrow Cove [SW 894 768] where there are green or pale brown, basic lithic tuff and acid tuff beds up to 1 m thick. These beds show evidence of reworking, channelling, and normal and inverse grading. The lenticular shapes of the tuff exposures inland on the southern limb are in part due to folding, but mostly they are a primary depositional feature. Some of the thicker tuffs can be seen to channel into the underlying sediments. Lateral variability also indicates that these rocks are localised expressions of volcanic activity.

Between the coast and Wadebridge, the Trevose Slate contains sporadic lenses of volcanic rock forming prominent grassy knolls.

Vesicular basalt, seen for example near St Merryn [SW 8950 7353; SW 897 728], forms a small proportion of the suite; massive hyaloclastite and tuffs predominate. The massive hyaloclastites, for example those exposed south-west and south-east of St Merryn [SW 8655 7282; SW 9027 7317], typically comprise small irregular cuspate fragments (up to 2 mm) of vesicular glass, now represented by cryptocrystalline and microcrystalline chlorite [E 62795], with a small proportion of albitic feldspar microphenocrysts. Fragment boundaries and vesicle walls are commonly coated with leucoxene, and vesicles are filled with chlorite or cloudy albite. Secondary albite crystals up to 3 mm across, also form patches of mosaic showing sericitisation. Massive hyaloclastites, grading into very fine-grained tuffs, are exposed in some small roadside cuttings and quarries, for example south and south-east of St Merryn [SW 8875 7282; SW 9000 7313], but coarse lithic crystal tuffs are only represented in field brash. The coarse-grained tuffs, for example at St Merryn [SW 8870 7371; SW 9020 7332], contain a variable proportion of lithic fragments; grey siltstone or silty mudstone clasts, up to 0.03 m across, predominate over fine basaltic clasts, and there is a high proportion of feldspar xenocrysts. The character of the more thinly bedded rocks and their association with the massive hyaloclastite suggest that they are the products of convective reworking and turbid flow on and around the piles of quenched basaltic glass, and could be more accurately referred to as bedded hyaloclastite. The coarser-grained tuffs are possibly the products of separate events, such as vent clearance and/or mass flow.

Between the Allen Valley Fault and the coast at Port Quin and Port Isaac bays, the Trevose Slate crops out in a broad east–west tract in the northern limb of the St Minver Synclinorium. This sequence includes the rocks formerly identified as the Pentire Pillow Lava Group (Gauss and House, 1972), a succession of pillow lavas and volcaniclastic sediments with dolerites. The pillow lavas, lava flows and intrusions are described in detail in Chapter three; the petrographically and geochemically similar volcaniclastic rocks are described here.

On Pentire Head and eastwards along the coast from Sandinway Beach [SW 9347 8094] to Sandy Bay [SW 9366 8060] and at Downgate Cove [SW 9790 8112], the volcaniclastic units occur interleaved with the extrusive volcanic rocks. These localities may have been close to volcanic centres because the lithologies present include unstructured conglomerates and pillow breccias, which are here interpreted as a proximal volcaniclastic facies. Hyaloclastites, formed by the rapid explosive quenching of lava flows, occur in the Sandinway Beach to Sandy Bay sections, north-east of Pennant Cottages [SX 0042 7901] and near Roscarrock [SW 9872 8095].

Some of the thinner volcaniclastic units, irregularly distributed in the Trevose Slate throughout much of the area, show graded bedding and are interpreted as volcanogenic turbidites rather than ash-fall tuffs. These epiclastites suggest a more distal position from the volcanic centres, although small bodies of basaltic lava occur sporadically throughout the Trevose Slate.

Plate 1 Palynomorphs of the district. All specimens figured below are preserved as negative images and housed in the Palaeontological collections of the British Geological Survey, Keyworth, Nottingham, UK. For each specimen, the name, author, sample locality, size, BGS specimen number, and position within the negatives catalogue are given.

a. *Lophozonotriletes lebedianensis* Naumova (1953). Harbour Cove Slate Formation [SW 9313 7980]. Max. diameter, 52 microns. MPK 9706 (162C5).

b. *Emphanisporites rotatus* (McGregor) McGregor (1973). Trevose Slate Formation [SX 1032 6812]. Max. diameter, 35 microns. MPK 9707 (215C1).

c. *Dictyotriletes emsiensis* [Allen] McGregor [1973]. Staddon Grit Formation [SX 0489 6775]. Max. diameter, 48 microns. MPK 9708 (211A3).

d. *Dibolisporites albitiensis* McGregor and Camfield [1976]. Staddon Grit Formation [SX 0535 6835]. Max. diameter, 32 microns. MPK 9709 (213A1).

e. *?Geminospora lemurata* (Balme) Playford (1983). Trevose Slate Formation [SX 0020 8098]. Max. diameter, 33 microns. MPK 9710 (215F3).

f. *Geminospora?*—monolete variant —possibly of *G. lemurata* (Balme) Playford (1983). Trevose Slate Formation [SX 0020 8098]. Max. diameter, 43 microns. MPK 9711 (219E6).

g. *Velamisporites perinatus* (Hughes and Playford) Playford (1971). Harbour Cove Slate Formation [SW 9313 7980]. Max. diameter, 30 microns. MPK 9712 (163E5).

h. *Camptotriletes paprothii* Higgs and Streel (1984). Harbour Cove Slate Formation (Gravel Caverns Conglomerate) [SW 9316 7979]. Max. diameter, 36 microns. MPK 9713 (166B3).

i. *Solisphaeridium* cf. *inaffectum* Playford (1981). Harbour Cove Slate Formation [SW 9313 7980]. Max. diameter of vesicle, 17 microns. MPK 9714 (162A5).

j. *Gorgonisphaeridium condensum* Playford (1981). Harbour Cove Slate Formation [SW 9313 7980]. Max. diameter of vesicle, 33 microns. MPK 9715 (163C3).

k. *Lophosphaeridium sp. 1.* Trevose Slate Formation [SW 9369 8056]. Max. diameter of vesicle, 38 microns. MPK 9716 (224A1).

l. *Veryhachium valiente* Cramer (1964). Trevose Slate Formation [SW 9369 8056]. Max. diameter of vesicle, 18 microns. MPK 9717 (223F1).

Plate 2a
A southerly view from Porthmissen Bridge [SW 8920 7645] of Marble Cliff.

An inverted succession of limestone and dark grey slaty mudstone beds of Givetian–Frasnian age, comprising the Marble Cliff Member, dips gently south-westwards. The limestone beds have a thickness up to 1 m, are graded and consist mainly of crinoidal debris. Gentle flexures occur in the vicinity of small faults and localised zones of tighter folding are due to synsedimentary slumping. (A15406)

Plate 2b   Gravel Caverns Conglomerate [SW9317 7978], north of Pentireglaze Haven.

Erosive base of polymict, matrix-supported, conglomerate in Gravel Caverns Conglomerate Member. Moderate to steeply northward-dipping, slaty cleavage, locally marked by pressure solution seams, shows refraction through the conglomerate. Elongate pebbles are partially rotated into the cleavage. S_{2n} cleavage (to the right of coin) dips gently southwards. Viewed from the west. (A15397)

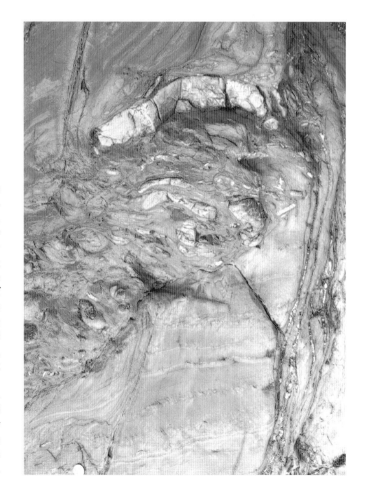

Marble Cliff Limestone Member

At Marble Cliff [SW 891 764] north of Trevone, a series of crinoidal turbiditic limestones, estimated by Mouravieff (1977) to be more than 80 m thick, have yielded extensive conodont faunas. These were first identified by House (1961), and have been described by Kirchgasser (1970), House et al. (1978) and Mouravieff (1977). At the type locality, Marble Cliff [SW 891 764], the beds are exposed in an inverted succession in a 300 m-long cliff section (Plate 2a) between Roundhole Point [SW 8925 7645] and Porthmissen Bridge [SW 8890 7630]. The member also crops out from Dinas Head [SW 8470 7615] to Barras Bay [SW 8585 7667] near Trevose Head, and from Little Cove [SW 8645 7595] to Cataclews Point [SW 8730 7612] west of Harlyn Bay.

The lithologies have been described by Gauss and House (1972) and by Beese (1984). A total of 159 limestone beds have been recorded within dark grey mudstones. The individual limestone beds, which range in thickness from 50 mm to 0.6 m, are dark bluish grey when fresh, but weather to a characteristic yellowish grey colour.

Tucker (1969) established the turbiditic origin of these limestones, noting sole marks including channelling, flute and groove casts and some tool marks, on about 40 per cent of the basal surfaces. Individual beds become finer grained stratigraphically upward, but no separate graded lower unit is recognised. The thicker beds commonly show laminated lower and upper units separated by a cross-laminated unit. At Marble Cliff, a pale greenish grey tuff in the middle of the section forms a useful marker band. The base of the member (Kirchgasser, 1970) is marked by a dolerite intrusion at the western end of Marble Cliff [SW 8891 7631] and the top is placed at the top of limestone 159 (Beese, 1984). Inland, the Marble Cliff Limestone is rarely seen, being represented only by thin turbidite developments on the foreshore at Rock [SX 395 755] and in a stream section at Gutt Bridge [SW 976 753]. Graded limestones within the dark grey slates of the Trevose Slate Formation in both the western [SW 932 813] and eastern [SW 9349 8106] cliffs of Rumps Point [SW 932 813] are of the same age.

Age

Stratigraphically useful faunas are uncommon in the lower part of the Trevose Slate. House (1963) identified goniatites, indicative of a late Eifelian age, at Booby's Bay [SW 853 758] and of an early Givetian age at Mother Ivey's Bay [SW 864 760] and Constantine Bay [SW 857 746]. This survey suggests that the Booby's Bay and Mother Ivey's Bay localities are at a similar stratigraphical level, and that the Constantine Bay locality is older. A limestone turbidite at Trevone Bay [SW 8908 7592] has yielded a lower *Polygnathus varcus* Biozone conodont fauna (House et al., 1978). At Pentonwarra Point [SW 890 760], ammonoids indicative of a Givetian age (*Maenioceras terebratum* Biozone) have been determined by House (1956, 1963). The mudstones at this stratigraphical level yield dacryoconarids and restricted horizons with gastropods, ammonoids, bivalves and brachiopods (Beese, 1982). Givetian conodonts (*Polygnathus varcus* Biozone) have been recorded from lenticular, graded limestones at Rock [SW 935 755] (Kirchgasser, 1970) and Rumps Point [SW 9322 8113] (upper *Schmidtognathus hermanni-Polygnathus cristatus* Biozone, Austin et al. 1992). The latter fauna belongs to the deep subtidal polygnathid biofacies.

The Marble Cliff Limestone has yielded a rich conodont fauna (Kirchgasser, 1970; Mouravieff, 1977; House, et al., 1978; Austin, et al., 1985) indicative of the upper *Schmidtognathus hermanni-Polygnathus* cristatus to the lower *polygnathus asymmetricus* biozones (Figure 5). The Givetian/Frasnian boundary is placed about 4 m below the top of the member (Kirchgasser *in* House and Dineley, 1985).

At Marble Cliff, the limestone beds (Beds 162–196 of Figure 6) above the Marble Cliff Member have yielded Frasnian (lower *Polygnathus asymmetricus* Biozone) conodonts. Conodonts from limestones included in megaclasts at the base of the Pentire pillow lava sequence at Com Head [SW 9400 8050] (Austin et al., 1992) give a Frasnian, post-lower *Ancyrognathus triangularis* Biozone age: the youngest unequivocally referred to the Trevose Slate.

In the vicinity of Helsbury Quarry [SX 088 791], grey micaceous sandstones have yielded a shelly fauna, including phacopid trilobites and stropheodontid brachiopods. A similar, but fragmentary brachiopod fauna has been recovered from a roadside exposure near Knightsmill [SX 0769 8051]. The presence of *Strophonelloides reversa* (Hall) indicates a late Devonian age.

Environment

The pervasive feature of the Trevose Slate is the regular lamination. Similar lamination has been reported from marine sequences in present-day and ancient basins; the Santa Barbara Basin in offshore California provides a close analogue to the Trevose Slate. There, the laminated sediments (Hulseman and Emery, 1961) are hemi-pelagic in origin, reflecting direct terriginous input under conditions of little or no current activity in anoxic bottom conditions in a barred basin. Successive lamina couplets are interpreted by Hulsemann and Emery (1961) as annual deposits, the coarser silty laminae representing increased seasonal runoff. A reported systematic variation in lamina thickness with depth, presumed to reflect longer-term climatic change, compares with the lamina bundles in the Trevose Slate. Laminated sediments are being deposited at the present day in the Santa Barbara Basin in water depths between 500 and 600 m. Interbedded with the laminated silty mudstones there are thin, slightly coarser beds showing crude normal grading which are interpreted as muddy turbidites. Also present are units, less than 1 m thick, of poorly laminated or homogeneous mudstone. These contain abundant burrows and remains of benthic organisms, indicating that they were deposited in periods of oxic bottom conditions. Hulsemann and Emery (1961) ascribe these oxygenated phases to the temporary influx of oxygenated water, possibly the result of storm activity. Similar beds, associated with scours and bioclastic debris in the upper part of the Trevose Slate sequence, appear to indicate deposition within reach of storm waves, and thus possibly in relatively shallow water.

Beese (1982) concluded that the bulk of the succession is of turbiditic origin because local alternations of silt and mud laminae occur in turbidite sequences on a scale not dissimilar to that seen in the Trevose Slate. The formation, however, does not contain amalgamation of laminae couplets, composite or multiple grading and wide variation in the thicknesses of the mudstone units commonly associated with turbidites. Despite the apparent turbidite characteristics of the sporadic beds of siltstone, the turbidite model seems to be inappropriate for the generation of the bulk of the finely laminated part of the formation.

The Trevose Slate thus appears to represent dominantly hemipelagic sedimentation in a barred marine basin with a tendency for bottom water anoxia. Under such conditions, varve-like laminae accumulated without disruption by benthic organisms. Phases of bottom-water oxygenation are represented by bioturbated units, prominent in the lower half of the formation. Superficially similar units, in the upper part of the sequence, display features indicative of deposition under the influence of wave action, suggesting increased oxygenation resulting from shoaling.

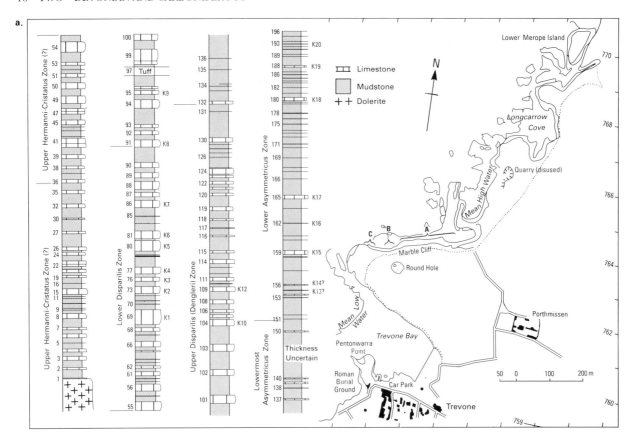

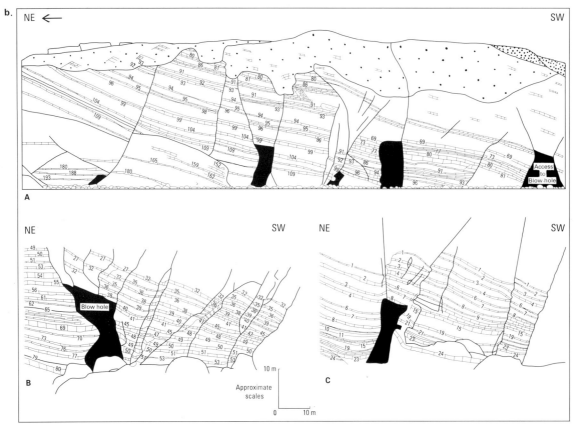

Figure 6 The type section of the Marble Cliff Limestone Member at Marble Cliff, Trevone, as figured by House et al. (1978). The top of the member is placed at the top of their Bed 159.

a. Location of Marble Cliff and a graphic representation of the sequence.
b. Sections of Marble Cliff, indicated in a), with numbered turbidite limestone beds.

The great thickness of the sequence indicates rapid subsidence and filling, with instability represented by small-scale slumping and the fine-grained turbidites. The thicker limestone turbidites of the Marble Cliff Limestone, comprise fossils and carbonate debris from a sublittoral environment (Tucker, 1969), which may have been derived from the basin bar, lying to the south-west, during a period of relative shallowing. The conodont faunas, which show no evidence of vertical reworking, show a mixture of shallow- and deeper-water forms (Mouravieff, 1977). The tuffs and agglomerates of the upper part of the formation give evidence of turbidity current activity in a basinal environment. Conodonts from a limestone megaclast included at the base of the pillow lava sequence indicate a mixed biofacies, representing redeposited debris in a peri-reefal setting (Austin et al., 1992). These volcanic sediments could have been derived from the northern margin of the basin, where more intense volcanic activity within the Trevose Slate suggests increasing crustal instability.

Harbour Cove Slate Formation

The Harbour Cove Slates were defined by Gauss and House (1972) as the lower part of the Upper Devonian purple and green slates of Reid et al. (1910). Beese (1984) later extended the formation to include the lithologically similar Merope Island Beds (House, 1961), and designated the cliffs between St George's Cove [SW 9217 7640] and Harbour Cove [SW 913 770] as the type section.

The formation consists of homogeneous green and grey-green slaty mudstones with thin beds of black, commonly pyritous mudstone and yellowish brown sideritic mudstone throughout. At the base of the formation, about 40 m of grey and green slates with variable colour banding (the Merope Island Beds) contains dark grey pyritous horizons with ammonoids.

The higher parts of the formation contain beds of purple mudstone, commonly with small-scale colour banding. Coarse-grained tuffs, up to 100 m thick, occur locally. Beds of ferruginous and calcareous nodules become increasingly common towards the top of the formation. This wide variation in slate colour and lithology is characteristic of the Harbour Cove Slate. Bioturbation is locally abundant, and minor slumping is sporadically developed.

The succession is particularly well displayed along the western side of the Camel Estuary, and a detailed log of the section is included in Gauss and House (1972, table 1, pp. 163–164). The stratigraphical base of the formation was defined by Gauss and House (1972) in the cliff facing Lower Merope Island [SW 8943 7700]. It is marked by the abrupt upward change from grey to green mudstone, a few metres above two beds of lava which lie within the Trevose Slate.

The lower boundary of the formation extends eastward from there across the Camel Estuary and is thrust-faulted and displaced by cross-cutting vertical faults. This thrust is offset by a north-east-trending fault at Chapel Amble [SW 982 753] and south-west of this locality an outlier of the formation occupies a fold core in the Trevose Slate. The base of the formation has been mapped as a conformable junction throughout the remainder of the St Minver Synclinorium, from St Tudy and Trelights to the coast at Polzeath. Beese (1984) estimated the Harbour Cove Slate to be 260 m thick near Padstow where it is in faulted contact with the succeeding Polzeath Formation.

East of the Allen Valley Fault, the Harbour Cove Slate is steeply dipping and has a sinuous outcrop, trending roughly north–south. The thickness is estimated to be about 600 m. Both the lower and upper boundaries are gradational, but neither is exposed. The base of the formation has been taken at the appearance of the characteristic green and greenish grey mudstones. No fauna has been recorded.

The formation also crops out in the inverted northern limb of the St Minver Synclinorium, where it includes a local facies variant, referred to as the Pentire Slate and Gravel Caverns Conglomerate by Gauss and House (1972). On the south west side of Pentire headland, the formation consists of grey-green mudstone with sporadic laminae and thin beds of siltstone and fine-grained sandstone, and thin decalcified beds and tuffs. Most of the sandstones are structureless, but some show cross-bedding. The scarcity of bioturbated horizons and the presence of many planar slump beds suggest relatively rapid sedimentation. At the south eastern limit of the coastal section, this slaty mudstone sequence is in faulted contact with the Gravel Caverns Conglomerate. Rare lenticular conglomerates within the slaty mudstone suggests a gradational conformable junction was originally present.

Gauss and House (1972) have estimated the thickness of the Harbour Cove Slate (their Gravel Caverns Conglomerate and Pentire Slate) to be about 430 m in this incomplete coastal section.

Inland, the Harbour Cove Slate outcrop runs eastwards from Port Quin Bay [SW 958 799] through St Endellion to the Allen Valley Fault. It is shown on the map to pass conformably southwards into the Polzeath Slate.

The Harbour Cove Slate also crops out in coastal sections extending between Varley Head [SW 985 815], Lobber Point [SW 992 811] and Portgaverne [SX 001 811]. At the last locality, a conglomeratic bed 5 to 15 cm thick in the Harbour Cove Slate is weakly channelled into dark grey, laminated Trevose Slate. The conglomerate consists of poorly sorted clasts of limestone, chert, mudstone and vesicular lava, set in a matrix of mudstone tuffite. Inland from here, volcanic rocks persist at this horizon and the contact with the Trevose Slate has been mapped as a thrust fault. Purple and green slates characteristically alternate in this inland outcrop of the Harbour Cove Slate although at Lobber Point, grey and silty green mudstones are exposed.

GRAVEL CAVERNS CONGLOMERATE MEMBER

The Gravel Caverns Conglomerate is here distinguished as a member in the highest part of the Harbour Cove Slate. The dominant lithology is greyish green mudstone similar to that at the type locality of the Harbour Cove Slate, with thin black, calcareous beds and, diagnostically, beds of poorly sorted, matrix-supported conglomerate up to 1.6 m thick (Plate 2b). The conglomeratic beds contain clasts of well-rounded siliceous mudstone, limestone, quartz-veined conglomerate and a variety of locally derived igneous rocks. Soft-sediment deformation is present in the softer locally derived clasts. The relationships of the conglomerates to the enclosing mudrocks are complex. Sheets and small channels occur commonly, and there is considerable invasion of the enclosing mudrock by pebbles at both upper and lower contacts. Significant dewatering processes appear to have operated soon after deposition. The imbrication of pebbles within the conglomerates is a tectonic phenomenon; the clasts having been rotated to render their long axes parallel to the slaty cleavage of the early deformation. Some interbedded sandstones are graded, fining upwards into rippled siltstones.

The Gravel Caverns Conglomerate is the product of tectonic instability in the sediment source area. The polymict nature and rounding of the pebbles suggest a geologically varied source area, with rounding taking place in an active environment, possibly a beach, before the material was released to the basin. The high matrix content, fluidisation, sheeted structure, and lack of good channel forms suggest a different final transport phase, possibly slumping. In an active volcanic setting, slopes may be steepened and there could be repeated detachment, sliding and fragmentation of lithified and partly lithified sequences. The sheeted conglomerates may represent debris flows and the channelled conglomerates could have been deposited from high-density turbidity currents. Individual components in the conglomerates could have undergone significantly different transport histories. Derived faunas give limited supporting evidence for the shoaling also deduced from volcanism. The occurrence of *Pachypora* spp. (Fox, 1905), and limestone clasts yielding Givetian conondonts suggest the source areas may have included Middle Devonian carbonate build-ups, possibly on volcanic highs.

In the coastal sections, the Gravel Caverns Conglomerate appears at the top of the Harbour Cove Slate. The contact with the Polzeath Slate [SW 935 795] is marked by the Polzeath Thrust (Selwood and Thomas, 1988), a structure which displays a zone of intensive deformation extending through up to 20 m of strata. Large-scale, fault-bounded lenses of dolerite and of the adjacent formations, are interleaved in slates much degraded by pervasive ductile thrust movements. This fault can be traced north-eastwards across the Pentire headland to emerge in the cliffs south of Carnweather Point [SW 951 802]. The regional mapping of the St Minver Synclinorium has shown that the Polzeath Slate overlies the Harbour Cove Slate comformably, but this relationship is unclear at many localities because of faulting.

The abundant pelagic faunas of the Harbour Cove Slate of the type area (Beese, 1984) include ammonoids and ostracods in dark grey and purple mudstones and conodonts in lenticular limestones. Taken together, they indicate a mid to late Frasnian age for the formation in the southern limb of the St Minver Synclinorium; an age constrained by Upper Frasnian ostracods from the overlying Polzeath Slate (Gooday, 1973). Gauss and House (1972) noted a significant upward change in the ammonoid faunas from *Ponticeras pedderi* to *P. prumiense* within the *Manticoceras cordatum* Biozone. Although bioturbated horizons occur locally in the formation, benthonic faunas are uncommon. Phacopid trilobites have been collected from the upper part of the formation at Daymer Bay [SW 927 776] (Thomas, 1909). In the northern limb of the St Minver Synclinorium, the Gravel Caverns Conglomerate has yielded derived pyritised Frasnian ammonoids (*M. cordatum* Biozone) (House et al., 1977). Conodont (Austin et al., 1992) and palynomorph (Selwood et al., 1993) assemblages indicate that the formation is here not older than Famennian. Marked diachroniety across the synclinorium is indicated by these faunal ages.

The Harbour Cove Slate was deposited in a marine basin with quiet sedimentation interrupted only by occasional volcanic activity. The latter may have been responsible for the triggering of local synsedimentary slumping. Oxygenation of the sea floor appears to have been irregular, with evidence (the black slates) of recurrent periods of euxinic conditions. The oldest purple mudstones recorded in the Trevone Basin, precursors of the purple mudstones of the Polzeath Slate occur in the highest part of the Harbour Cove Slate. The depositional significance of the colour change is not understood. Mossbauer spectroscopy studies indicate that although both the green and purple mudstones contain ferrous and ferric iron within the clay-mineral lattices, only the purple slate contains coatings of ferric iron on the surface of the particles. This difference probably reflects an early diagenetic change. Beese (1984, p.69) identified two facies which alternated throughout the Harbour Cove Slate. These are green mudstones with thin black, mudstone bands and interbedded fossiliferous dark grey and purple mudstones, which he attributed to 'inner basin' and 'off-fan' environments respectively.

Polzeath Slate Formation

The term Polzeath Slate was introduced by Gauss and House (1972) to describe the Upper Devonian purple and green slates of Reid et al. (1910) in the Trevone Basin. The formation was redefined and subdivided by Beese (1984) who restricted the type section to the cliffs between Daymer Bay [SW 9263 7777] and The Greenaway [SW 9285 7843]. His lithological subdivisions based on the colour of the mudstones and the character of the silty mudstones, could not be mapped inland.

The base of the formation which is not observed in continuous section, is taken where the colour changes from dominantly grey-green to purple in a gradational mudstone sequence. In the coastal sections, the contacts with the Harbour Cove Slate are faulted. Near Hawker's

Cove [SW 9130 7765] at Daymer Bay [SW 9263 7779] and at Pentire Headland, the Harbour Cove Slate is thrust over the Polzeath Slate.

Inland, the Polzeath Slate rests conformably on the Harbour Cove Slate. The thickness of the formation is difficult to determine due to multiphase deformation. Gauss and House (1972), and Beese (1984) estimated it to be 320 m and 240 m respectively.

The formation consists of cleaved, purple mudstones in which bedding is picked out by pale green and grey beds and laminae, and by thin sandstone and laminated siltstone horizons. Thicker beds of green mudstone occur in association with dark grey mudstone which is sporadically fossiliferous. Large volcanic bodies have been mapped in the easternmost parts of the outcrop area, but the only evidence of extrusive volcanic activity in the coastal exposures is thin chlorite-rich, tuffaceous, mudstone beds.

On the south side of Hayle Bay [SW 935 790 and SW 9320 7895] small dolomitised limestone nodules, possibly sideritic, occur closely packed in beds up to a few metres thick in purple mudstones. Primary nodules are present, but many are thin lenticular beds of carbonate which have been fractured and rotated in the cleavage.

Sedimentary structures in thin sandstones include graded bedding, load structures and rare cross-lamination. The mudstones are locally bioturbated. Secondary alteration of the purple coloration to green is extensively developed; the colour contacts are diffuse, and commonly cut across the bedding.

The Polzeath Slate has yielded little fauna or flora other than pelagic ostracods which are particularly abundant at a few, mostly micaceous, horizons in the purplish dark grey mudstone. Except where bedding and cleavage are broadly parallel, most of the fossils are indeterminate. Gooday (1973) and Beese (1984) have recorded localities indicating that the formation spans late Frasnian to mid-Famennian times.

The Polzeath Slate was deposited in a basin far removed from sources of coarse sediment. The intercalation of laminated silty mudstones and dark grey mudstones has been interpreted by Beese (1984) as indicative of recurrent prograding outer (turbidite) fan and off-fan environments respectively. The nodular carbonate beds suggest that seafloor rises, possibly of volcanic origin, were present within the basin. Such rises could have made some local contribution to sediment supply. Any volcanic contribution to the sequence was, however, minimal.

TINTAGEL SUCCESSION

The Tintagel Succession (Figure 3) was first recognised by Parkinson (1903) and modified by Dewey (1909). They divided the thick succession of greenish grey slates into formations on the basis of their inferred stratigraphical position, mineralogy and lithology. These formations are difficult to distinguish in the field, and subsequent mapping revealed conflicting interpretations (cf. Dewey, 1909; Wilson, 1951; Batstone, 1959; Freshney et al., 1972). In the Boscastle (Sheet 322) district, Freshney et al. (1972) recognised a Devonian greenish grey slate

formation, the Tredorn Slate, and three Carboniferous formations, the Barras Nose, Tintagel Volcanic and Trambley Cove formations. Subtle lithological variations occur in these formations, some of which are emphasised by low-grade metamorphic mineralogy, but most of these minor variations are not mappable. In this survey, all the Devonian greenish grey slates are referred to the Tredorn Slate, a practice adopted by most recent workers in the area (e.g. Hobson and Sanderson, 1975; Selwood and Thomas, 1986a; Warr, 1988). The economically important Delabole Member is a lenticular body within the Tredorn Slate.

Facies analysis of the Tintagel Succession suggests a rapid change from outer-shelf, muddy deposition in the late Devonian, to fully basinal, black shales with euxinic bottom conditions in the early Carboniferous. A within-plate tectonic setting for the Tintagel Succession is indicated by the chemistry of the Tintagel Volcanic Formation (Robinson and Sexton, 1987).

Tectonically the succession includes the Tintagel High Strain Zone in which illite crystallinity studies (e.g. Brazier et al., 1979) have revealed epizone metamorphism. Selwood and Thomas (1986a) included the Tintagel Succession in their Tredorn Nappe.

Formations of the Tintagel Succession are represented in the northern part of the district. They extend westwards from the north-western margin of the Bodmin Moor Granite, in a broad sweep around the Davidstow Anticline, to appear in coastal sections extending as far south as Jacket's Point [SX 033 831]. There the succession is thrust on a low-angle fault over beds belonging to the Bounds Cliff Succession. Inland, the Tintagel Succession is overthrust on high angle faults by the Padstow Succession.

Tredorn Slate Formation

The original usage of the Tredorn Beds (Parkinson, 1903) is here extended to include within the Tredorn Slate all greenish grey slates in the Tintagel Succession. They include the Upper Delabole Slate of Freshney et al. (1972), the Woolgarden Phyllites of Dewey (1909), and the Tredorn Beds of Parkinson, (1903).

The Tredorn Slate consists of greenish grey quartz-chlorite-mica slate. At the type locality (Simpson, 1959), Tredorn [SX 107 897] south-east of Boscastle, the formation is poorly exposed, but the coastal section between Saddle Rocks [SX 0741 9030] and California Quarry [SX 0902 9084] provides a good reference section and includes the upper boundary of the formation at California Quarry. The slates are locally interbedded with thinly bedded, commonly lenticular, bioclastic limestone and dolomite beds, up to 0.15 m thick, and with sandstone, siltstone and rare tuff beds. Slate and siltstone are interlaminated in places, and in the metamorphic aureole of the Bodmin Moor granite this gave rise to distinctively banded slates referred to as the 'Woolgarden Phyllites'.

Rust-spotting, derived from the oxidation of pyrite, and spotting caused by low-grade metamorphism, are extensively developed throughout the Tredorn Slate.

Porphyroblasts of orthoclase and chloritoid minerals are obvious in hand sample, and the orthoclase may be so abundant as to coarsen the texture of the slate. Under the microscope, the complex fabrics reflect a multiphase deformation history (see Chapter six).

The southern boundary of the Tredorn Slate is marked by vertical or steeply north-dipping faults. To the west an overthrust relationship with the Jacket's Point Slate has been identified.

The thickness of The Tredorn Slate is difficult to calculate because of a lack of marker horizons and the occurrence of multiphase deformation, which includes recumbent isoclinal folding. A structural thickness of 500–800 m has been estimated.

Faunas are unevenly distributed through the formation. In the slates, macrofossils are largely restricted to thin horizons rich in spiriferid and rhynchonellid brachiopods, polyzoa and crinoid ossicles. Species diversity is low, but the number of individuals is characteristically high. Most fossils are preserved as indeterminate, flattened, chloritic films. The Upper Devonian brachiopod *Cyrtospirifer verneuili* (Murchison), popularly known as the 'Delabole butterfly', is well represented.

Stage-diagnostic fossils have not been found in the district, but thin, partly dolomitised limestones at Trebarwith Strand [SX 049 871] immediately north of the district boundary, have yielded conodont faunas indicative of a Famennian upper *Palmatolepis marginifera* to lower *velifer* Biozone age (Stewart, 1981), (the *scaphignathus velifer* Zone is no longer used and the *P. marginifera* Zone now includes this shallow water biozone).

A general absence of soft sediment burrowing, and the thick, generally barren mudstones suggest rapid deposition. It seems that opportunistic species were only able to colonise the sea floor infrequently, as conditions temporarily ameliorated. The thin, fine-grained sandstones and lenticular limestones contained within the sequence appear to be distal turbidites, possibly introduced as a result of storm action. Analysis of the conodont faunas (Stewart, 1981) suggests an outer-margin, shelf environment with some displaced shallow-water species; an interpretation consistent with the overall facies of the formation.

DELABOLE MEMBER

The Delabole Member is here defined as the bluish grey slate that has a lenticular mapped form within the greenish grey and grey-green slates of the Tredorn Slate Formation. It can be traced in an arcuate outcrop around the Davidstow Anticline, and is represented in a faulted slice south of Trebarwith. The member equates with the Lower Delabole Slate of Freshney et al. (1972). The type locality is at the Delabole Slate Quarry [SX 075 839] where it is exposed in the lowest part of the quarry. At the type section, it consists of a well-cleaved, very fine-grained, bluish grey, chlorite- muscovite slate that is hard, close-textured and commonly with a silky lustre. Pyrite cubes and ovoids after pyrite are common, and white mica and chlorite may give a speckled appearance to the slates. Thin silty laminations of paler colour than the enclosing slate are locally developed; at the type locality the slaty cleavage is parallel to bedding.

Flattened, chlorite-coated specimens of *Cyrtospirifer verneuili*, are common at some levels.

The member is poorly exposed away from the quarry and although everywhere it is predominantly grey slate, it also includes greenish grey, less finely cleaved slates. Both the lower and upper boundaries with the Tredorn Slate are thought to be gradational. One of the boundaries, is exposed in the north-east part of Delabole Slate Quarry; this is probably the upper boundary although the direction of younging is unknown. A structural thickness of 30 to 50 m is estimated for the member.

As with the Tredorn Slate, an outer shelf environment is proposed, but the dark colour of the slate and abundance of pyrite suggest deposition involving local anoxic conditions.

Barras Nose Formation

The Barras Nose Formation (Freshney et al., 1972) consists mainly of dark grey to black, pyritous, cleaved mudstones with laminated siltstone horizons and interbedded thin tuff. Thin lenticular limestones and calcareous laminae are developed near the top of the formation. Horizons of thin-shelled posidoniid bivalves and local bioturbated beds have also been recorded.

The base of the Barras Nose Formation is everywhere faulted in the district; in the Boscastle (Sheet 322) district the formation rests conformably on the Tredorn Slate (Freshney et al., 1972). A faulted junction is exposed at Tregardock Beach [SX 041 843], where the Barras Nose Formation is faulted against Tredorn Slate and Jacket's Point Formation. North-west of Bodmin Moor, [SX 179 851] in a similar structural setting, the Barras Nose Formation is reduced to a fault-bounded lens beneath the Tintagel Volcanic Formation.

A stratigraphical thickness of over 24 m was estimated by Freshney et al. (1972); and Warr (1991) has calculated a structural thickness of 80 to 100 m. The only reliable dating comes from conodonts obtained from limestones at the top of the formation, which show ranges within the Tournaisian *Scaliognathus anchoralis-Doliognathus latus* to *Gnathodus texanus* biozones. At the northern limit of the district, Stewart (1981) recorded a *Gnathodus texanus* Biozone fauna from white saccharoidal limestones in the Trebarwith Strand valley [SX 0554 8652].

Facies analysis (Selwood and Thomas, 1986b) suggests that deposition occurred within a marine, deep-water environment, mainly under euxinic conditions.

Tintagel Volcanic Formation

The Tintagel Volcanic Formation (Freshney et al., 1972), formerly the Volcanic Series of Parkinson (1903), consists of highly altered volcanic rocks including poorly sorted, fine- to coarse-grained calcareous tuffs, agglomerates and subordinate recrystallised carbonate-rich lavas. No recognisable pillow lava is preserved, but some of the basalt clasts in the agglomerates exhibit quench textures and plagioclase microlites typical of rapidly chilled submarine lavas. The clasts are either plagioclase phyric or aphyric glassy basalt, with minor holocrystalline microdolerites. Also present are rare interbedded lenticular beds of pink limestone and green and pale grey slate including volcanic bombs and lenses of tuff.

The agglomerates are mainly composed of rounded, flattened fragments of dark grey lava characterised by abundant pink calcite amygdales (Plate 3a). Long axes of clasts show a pronounced north-north-west-trending orientation which is parallel to the tectonic transport direction. The tuffs are characteristically greenish grey in colour, commonly banded and may include volcanic bombs. The formation records a major pulse of explosive volcanic activity in which many of the tuffs could have a reworked hyaloclastite origin.

The volcanic sequence has been profoundly altered by both deformation and regional metamorphism to lower greenschist facies (biotite zone). Chlorite, the principal metamorphic phase, formed with the growth of a penetrative cleavage which is particularly well-developed in the finer volcaniclastic rocks.

A section in the Trebarwith Strand valley [SX 053 865], which features the conformable basal contact with the Barras Nose Formation, has been described by Freshney et al. (1972, pp.42–43). Agglomerates in the higher parts of the sequence are exposed on the Trebarwith Strand foreshore [SX 048 865]. Although all lithological units are lenticular, a structural thickness ranging up to 90 m can be estimated.

The age of the Tintagel Volcanic Formation is not known. Conodonts of the *G. texanus* Biozone (Tournaisian) have been obtained from the top of the underlying Barras Nose Formation. Bleached early Carboniferous conodonts have been reported (Stewart, 1983) from recrystallised pink limestone within the volcanic sequence.

Trambley Cove Formation

The Trambley Cove Formation rests conformably on the Tintagel Volcanic Formation. The top of the formation is not seen in the district, being everywhere faulted against the Tredorn Slate. Freshney et al. (1972) suggested a thickness of 18m for the Trambley Cove Formation of the Trebarwith area.

The Trambley Cove Formation (Dewey, 1909), consists of finely banded, pale to dark grey and black, cleaved mudstones with thin beds of commonly cross-bedded siltstone and greenish brown tuffaceous beds. The formation has been described in detail by Freshney et al. (1972) and McKeown et al. (1973). Silicification of the mudstones and siltsones to form thinly bedded cherts is common, particularly in areas of intense deformation. In inland exposures, the slates weather to a pale greenish grey, and may show a dark grey variegation.

No fauna has been recorded from the formation in the district, but a late Dinantian to early Namurian age has been suggested for outcrops in the adjacent Tavistock district (Stewart, 1983). Facies analysis suggests that the environments of deposition in the Trambley Cove Formation were a continuation of the basinal marine conditions of the Barras Nose Formation.

BOUNDS CLIFF SUCCESSION

The Bounds Cliff Succession occupies a low structural level; cropping out between Port Isaac Bay and Westdowns. It is faulted to the north and north east against the Tredorn Slate of the Tintagel Succession, and is overthrust to the south east by the Trevose Slate of the Padstow Succession.

Two formations have been identified in an inverted succession. The Jacket's Point Slate Formation, an outer shelf/basin transition facies, and a stratigraphically overlying heterogeneous group of slaty mudstones, which have been correlated on the map with the Harbour Cove Slate.

Jacket's Point Slate Formation

The Jacket's Point Slate (Selwood and Thomas, 1986a) consists of more than 100 m of grey slaty mudstones with sandstones, conglomerates and associated volcanic rocks. The formation crops out in about 3 km of cliffs extending north from Bounds Cliff [SX 0178 8128]; the succession is inverted and overthrust by Tredorn Slate at its northern end. The junction with the Harbour Cove Slate, believed to be conformable, is exposed but complicated by faulting at the southern end of the section.

The formation is dominated by pale to mid-grey mudrocks which, apart from restricted bioturbated horizons, are characteristically featureless. Locally the slates are finely laminated, and at higher stratigraphical horizons interbanded purple slates characteristic of the succeeding formation appear.

The sandstones are irregularly interbedded with the slates. They are fine to medium grained, and show both sheet and lenticular geometries. Gutter casts filled with fine- to medium-grained, fairly clean quartz sandstone are common. Winnowed lag deposits, concentrated at the base of these fills, suggest that the cutting and filling were separate events. In places the infills show no internal structure, but more usually coarse lamination, crude grading, or more rarely, low-angled, small-scale cross-stratification occur. The sandstones may extend above and beyond the scour form, either wedging out rapidly in the slate or integrating into coeval sand sheets. Such gutter casts, although formed by quite separate erosive events, appear as sole structures on the bases of sheet sandstone beds; they may show a preferred orientation, but this varies with stratigraphical level.

Thin, uniform, sandstone sheets are common. They show sharp planar or gently undulatory bases, and graded or laminated tops, which only rarely carry low-amplitude wave ripples. Apart from infrequently bioturbated tops, faunas are absent.

The sheeted sandstones and gutter-cast sandstones appear to represent the result of isolated storm events which introduced sand from more marginal areas of the basin. Much of the clay material appears to have accumulated rapidly, with little time for bioturbation.

Intraformational, matrix-supported conglomerate is irregularly developed throughout the formation. It occurs in beds up to 2 m thick, bearing derived fossils,

Plate 3a Sheared tuff/volcanic breccia [SX 0470 8629], Port William, Trebarwith Strand.

Wave-worn surface exposing lava fragments of different types, elongate and flattened in the plane of cleavage, within the Tintagel Volcanic Formation. The surrounding matrix is fine- grained, with a slaty fabric composed of white mica and chlorite minerals.(A15392)

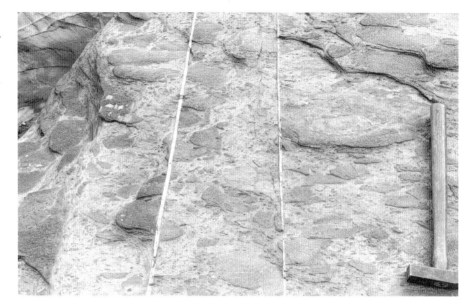

Plate 3b Old quarry face [SW 9319 8080], coastal footpath, 400 m south of Rumps Point.

Highly vesicular pillow lavas showing concentric arrangement of vesicles and axial cavities within individual pillows. A local flattening fabric is evident. Viewed from the north. (A15400)

Plate 3c Cliff exposure [SW 9349 8126] some 400 m east of Rumps Point.

Viscous flow structure at the top of a dolerite sill originally intruded into calcareous mudrocks, of the Middle Devonian Trevose Slate Formation, which have become extensively adinolised. This feature shows the section to be the right way up and gently inclined. (A15388)

and small carbonate and mudrock clasts; the clasts are of local derivation and of early diagenetic origin. The mudstone matrix accounts for more than half of the volume of these beds. Most conglomerate appears to have developed from local synsedimentary slides, and is broadly lenticular in form, but shows no obvious down-cutting relationships. Complete sedimentary dismemberment is revealed in a slump sheet at Bounds Cliff [SX 0304 8229], with a thickness approaching 3 m and a lateral extent of over 20 m. Sandstone beds are tightly and irregularly folded and the interbedded slates show numerous water escape structures, many cutting through bioturbated mudstones. Nodules of siderite and calcite, commonly enclosed in lenses of cone-in-cone calcite, are characteristic of the formation.

Deformed shelly faunas, mainly spiriferid and rhynchonellid brachiopods, occur in beds up to several metres thick; many of the beds are bioturbated, and separated by generally barren interbeds. At rare intervals, more diverse faunas are present and these include juvenile goniatites, orthocones, bivalves and gastropods. Poorly preserved goniatites from Barrett's Zawn [SX 0278 193], and north of Delabole Point [SX 0302 8230] indicate a Frasnian age.

Fine- to medium-grained, matrix-rich, lithic and crystal tuff with local agglomerate and grey and brown laminated slaty mudstone, appear in irregularly interbedded units up to 10 m thick [e.g. at Jacket's Point SX 0368 8338 and Barretts Zawn SX 0272 8187]. Soft sediment deformation is prominent, particularly where partially consolidated sediments were incorporated into and intruded by vesicular lava near the sedimentary surface. Locally, lavas appear to have fluidised the enclosing sediments and brecciated the overlying sediments.

At some localities, for example south of Crookmoyle Rock [SX 0314 8253], vesicular lava showing well-developed pillow structure occurs in tuffaceous slate; isolated flattened lava bodies appear to represent detached pillows.

The thick succession of rapidly deposited greenish grey mudstones that form the bulk of the Jacket's Point Slate are similar to those characteristic of the outer shelf environments elsewhere in the region. They are distinguished, however, by the presence of abundant storm-generated sandstone beds, slumps and intraformational conglomerates. The last two indicate sedimentary instability directly relating to the steepening of slopes, and are in places associated with volcanic rocks. A position high on the shelf/basin margin is indicated.

Harbour Cove Slate Formation

Although the junction of the Jacket's Point Slate and the Harbour Cove Slate is complicated by faulting at Bounds Cliff, some 500 m inland the boundary between the two formations has been interpreted as conformable. This view is reinforced by the appearance of beds of purple slate in the higher stratigraphical levels of the inverted Jacket's Point Slate in the coastal section. These purple slates have previously been placed in the Polzeath Slate (Hobson, 1975; Selwood and Thomas, 1986a; Warr,

1989). They are reassigned here to the Harbour Cove Slate on the basis of their lithological similarity. The Harbour Cove Slate in this area has suffered extensive tectonic dismemberment and pervasive recrystallisation, probably due to overthrusting by the Trevose Slate.

The coastal section has revealed foresets of symmetrical ripples with low amplitudes (10 mm), rounded crests and relatively long wavelengths. These recall spillover oscillation ripples, a feature of sandy tempestites (Seilacher, 1982). Thin tuffs and current-bedded sands have also been recorded; both have suffered dolomitic replacement to varying degrees. Poorly preserved pyritised ammonoids from the eastern part of Tartar Cove [SX 0108 8117] (Selwood and Thomas, 1986a) suggest an early Famennian age.

LISKEARD SUCCESSION

South of Bodmin Moor, Burton and Tanner (1986) demonstrated the existence of a persistent area of shelf deposition during mid to late Devonian times in the Liskeard area, between the South Devon and Trevone basins. This shelf was limited westwards by the St Teath–Portnadler Fault Zone, and eastwards by the Otterham Fault Zone (Selwood, 1990). The Liskeard Succession (Figure 3), above the Lower Devonian Staddon Grits, contains a mixed argillite sequence, in which formations are distinguished by the character of the matrix argillites and by various associations of sandstones, limestones and volcanic rocks. Open shallow-marine shelf conditions persisted from Emsian until early Upper Devonian times, when basinal conditions were established.

Rosenun Slate Formation

The Rosenun Slate Formation was introduced by Burton and Tanner (1986) to describe dull grey and black, commonly calcareous mudrocks in which decalcified horizons give a characteristic ochreous weathering. In the upper parts of the formation, a series of limestone beds and volcaniclastic sediments occur. The formation is richly fossiliferous and yields a trilobite-brachiopod fauna of late Eifelian to Givetian age. Open shelf conditions were maintained throughout deposition.

The Rosenun Slate Formation has been mapped (Burton and Tanner, 1986) to the south-east of the district boundary. The hornfelsed black and dark grey slates with lenticular calcsilicate and metavolcanic rocks, flanking the south and south-west margins of Bodmin Moor, represent its development within the metamorphic aureole of the granite. These beds are limited westwards by various north-west-trending faults in the Cardinham Fault Zone.

DEPOSITIONAL ENVIRONMENTS

The complex disposition of Variscan facies across central south-west England is most readily explained by intra-shelf basin development initiated on deep seated east–

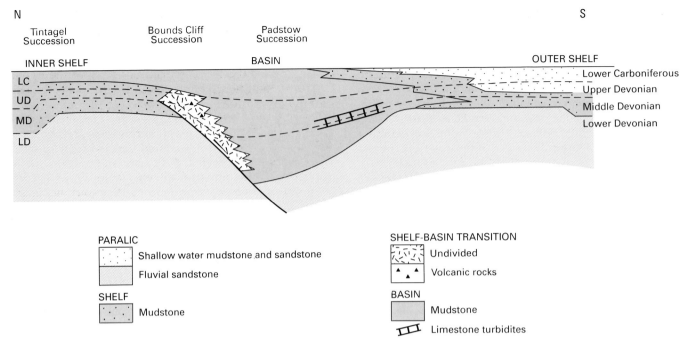

Figure 7 Half-graben facies model for the Trevone Basin.

west fractures in late Early Devonian times. A half-graben model (Figure 7) was proposed by Selwood and Thomas (1986b), and Selwood (1990) summarised evidence for the South Devon and Trevone basins, arranged along strike and separated by the Liskeard High. This high appears to have been defined by deep seated north-west-trending oblique, transfer faults, now expressed as the Cardinham–Portnadler and Otterham fault zones. Paralic, outer shelf, basinal and shelf-basin transition facies have been identified within the four successions represented in the district (Figure 7). Although largely isolated from one another tectonically, continuity of age-related facies usefully constrains the relative depositional setting of these successions. The Bounds Cliff Succession, currently represented at the lowest structural level, is taken as the datum.

The Bounds Cliff Succession lies in the footwall of a major thrust which carried the Padstow Succession northwards; the amount of displacement is unknown. A southern provenance for the Padstow Succession, relative to that of Bounds Cliff Succession, is consistent with its facies analysis which gives evidence of an active, fault-controlled basin in mid to late Devonian times. Concurrently the Bounds Cliff Succession shows a shelf to basin transition. A north–south passage between the two successions is likely; this view is supported by the spread of basinal purple and green slates across both areas in late Devonian times.

The paralic sediments which comprise the Staddon Grit and Meadfoot Group, at the base of the Padstow Succession, passed eastwards into shelf deposits in the Liskeard area (Burton and Tanner, 1986). Younger stratigraphical levels show contrasting sedimentary histories. On the Liskeard High, shelf facies continued

into the late Devonian, but in the coeval Trevone Basin deep-water conditions prevailed.

There is little indication that the outer shelf of the Trevone Basin carried a carbonate complex in the Middle to early Upper Devonian times. Rather, an onlap of clastic sediments from the Gramscatho Basin to the south is observed (Holder and Leveridge, 1986), and this probably supplied arenites to the Trevone Basin. Black shales indicate that the Trevone Basin was already developing in Emsian times. Following deposition of thick Middle Devonian mudstones a change in sedimentary regime is indicated about the Middle/Upper Devonian boundary by local limestone turbidites and volcanic activity. The distribution of these volcanic rocks in the basin directly reflects control of volcanic activity by extension along basement fractures.

Lateral passage from the Padstow Succession to the bounding shelf deposits of the Liskeard High are obscured by contact metamorphism by the Bodmin Moor granite, and by complex displacements along the Cardinham Fault Zone: the late expression of the controlling transfer fault.

The position of the Tintagel Succession, which is now thrust over the Bounds Cliff Succession, is problematical; it may have accumulated on the northern (inner) or southern (outer) shelf of the Trevone Basin.

ALTERNATIVE VIEWS ON THE STRATIGRAPHY OF THE PADSTOW SUCCESSION

Since the survey was completed Austin et al. (1992) and Selwood et al. (1993) have presented biostratigraphical evidence to support the introduction of a St Mabyn Succession, and the continued usage of the Pentire and Trevone successions. Within their scheme the separate identity of the Trevose Slate, Pentire Volcanic, and Pentire Slate formations is maintained and the Polzeath Slate is excluded from the Pentire Succession (Figure 8). The lithostratigraphical correlation of formations within the Padstow Succession made above, is disputed by these authors.

They argue that the Trevone and Pentire successions are separated by the Polzeath Thrust, in the structure here identified as the northern limb of the St Minver Synclinorium. They indicate that the Harbour Cove Slate represented on the 1:50 000 map and described in this memoir, includes parts of their Trevose Slate, Pentire Volcanic and Pentire Slate formations forming part of

their Pentire Succession, and parts of the Polzeath Slate and other formations ranging up into Lower Carboniferous cherts in the Trevone Succession (see Figures 8 and 40). Much of the Trevose Slate featured east of the Allen Valley Fault on the 1:50 000 map is referred by them to a newly distinguished St Tudy Slate Formation of Lower Carboniferous age.

Selwood and Thomas (1993) have also differentiated an area of California Slate Formation from the Trevose Slate represented on the 1:50 000 map north-west of Bodmin Moor. North of the district, this California Slate Formation overlies the Tredorn Slate. So extended, the Tintagel Succession equates broadly with the zone of epizone metamorphism identified by Warr and Robinson (1990).

Figure 8 Alternative nomenclatures for the Padstow Succession: A and B after Gauss and House (1972), C after Austin et al. (1992).

		A	B	C
Series	Stage	Trevone succession	Pentire succession	Pentire succession
LOWER CARBONIFEROUS	TOURNAISIAN			
UPPER DEVONIAN	FAMENNIAN	Polzeath Slates	Polzeath Slates	Gravel Caverns Conglomerate Member
				Pentire Slate Formation
	FRASNIAN	Harbour Cove Slates	Gravel Caverns Conglomerate	
		Merope Island Beds	Pentire Slates	Pentire Volcanic Formation
		Longcarrow Cove Beds	Pentire Pillow Lava Group	Trevose Slate Formation
		Marble Cliff Beds		
MIDDLE DEVONIAN	GIVETIAN	Trevose Slates	Trevose Slates	
	UPPER EIFELIAN			

THREE

Basic igneous rocks and associated metamorphism

The northern and external Rhenohercynian zone of the Variscan orogen is characterised throughout its length by extensive volcanism, mainly during the late Devonian to early Carboniferous periods. On the basis of their eruptive setting two main basaltic groups are generally recognised: those formed at a spreading axis with mid-ocean ridge chemical features, and those formed in an ensialic or within-plate setting with enriched chemical features. The basic rocks seen in north Cornwall are representative of this latter period of Variscan volcanism, and exemplify the diversity of composition exhibited by basaltic rocks erupted in ensialic basin settings.

In common with other basic rocks of the Variscan orogen the basic rocks of north Cornwall have undergone variable low-grade metamorphism, as well as deformation. The early terminology assigned to the volcanic rocks reflects this alteration, such that doleritic intrusions were referred to as greenstone, diabase, epidiorite and greenschist, amongst others, whereas basaltic lavas were generally called spilites (e.g. Dewey, 1914). None of the basic rocks within the area have escaped some degree of alteration, and are more strictly termed low-grade metadolerite, metabasalt and metabasic volcaniclastic rocks. The study of relict mineral assemblages, textures and stable-element chemistry enables their original composition to be determined, and as far as possible a nomenclature based on primary features is used throughout the text.

Three modes of emplacement can be recognised, each of which has given rise to different suites of volcanic rocks. Firstly, the deposition of volcanic fragments in water, together with some current reworking, has produced a suite of aquagene volcaniclastic rocks. Secondly, the submarine extrusion of basic lava onto the seafloor has produced mainly basaltic pillow lavas and minor, associated pillow breccias. Thirdly, dolerite sills and subordinate dykes were produced by the intrusion of basic magma into the sedimentary pile, often at a high level near the water–sediment interface.

The pillow lava at Pentire Point [SW 923 805], within the Harbour Cove Slate Formation, is Frasnian in age (Gauss and House, 1972), although inland to the east pillow lavas occur throughout the Upper Devonian (Reid and Dewey, 1908; Dewey and Flett, 1911) and within the Middle and Upper Devonian Trevose Slate Formation. Apart from the minor occurrence of tuffaceous beds in Middle and Upper Devonian rocks (Reid et al., 1910), the greatest development of volcaniclastic rocks is in the Tintagel Volcanic Formation [SX 047 892] in the Lower Carboniferous (Francis, 1970; Freshney et al., 1972) of probable Viséan age. Emplacement ages for some of the high-level intrusions that cause soft-sediment deformation features or have pillowed tops, can be inferred to be similar to the depositional ages of the enclosing sediments. These features suggest that the dolerite dykes at Dinas Head [SW 847 761] were intruded into the Trevose Slate during the late Middle Devonian, whereas the dolerite intrusions on either side of the Camel Estuary north of Padstow may have spanned both the Frasnian and Famennian.

Both extrusive and intrusive rocks appear to represent various intermittent volcanic episodes throughout the late Middle Devonian and Upper Devonian, followed by a brief cessation of activity before the Viséan event.

On the basis of their observed and inferred primary petrography, together with geochemical discrimination techniques, the basic volcanic rocks can be divided into the following suites:

Lower Carboniferous:
Tintagel Volcanic Formation.

Middle and Upper Devonian:
Tuffaceous volcaniclastic rocks (unnamed).
Pentire alkali basalt lava.
Anhydrous tholeiitic dolerite.
Anhydrous alkali dolerite.
Hydrous alkali dolerite.

The relationships between the lava and dolerite are rarely observable in the field. Geochemical data, however, can be used to link extrusive and intrusive rocks into comagmatic groups; the data suggest that the Pentire basaltic lava flows are related to the anhydrous alkali dolerite intrusions. Representative chemical analyses of lava and intrusive rocks are shown in Table 1.

MIDDLE AND UPPER DEVONIAN

Tuffaceous volcaniclastic rocks

This suite encompasses small, scattered outcrops of volcaniclastic rocks within mainly Middle Devonian sedimentary rocks (Figure 9).

Some outcrops have a generally basic composition and comprise crystal and lithic tuff and lapillistone. Crystal fragments are almost exclusively albite, which is thought to represent original calcic plagioclase replaced during low-grade metamorphism. Most of the lithic clasts are fine-grained aphyric and plagioclase-phyric basalt. Rarely, some volcaniclastic rocks have a more mixed nature and include epiclastic fragments derived from the locally enclosing sediments. Most of the finer-grained, in places laminated, tuff beds have been deposited through a passive water column, although there is also evidence for subsequent current reworking by mass-flow mechanisms.

Table 1 Representative analyses of different volcanic suites from north Cornwall.

Sample No.	1	2	3	4	5	6	7	8	9	10
Field No.	PP7R	KH3	WA2	PARK2	STE2	RU1	STS7	PA4	TV16	TV36
Major oxides (wt %)										
SiO_2	36.80	41.82	49.10	51.18	46.18	42.94	44.97	48.76	37.87	40.94
TiO_2	2.34	3.80	1.65	1.50	1.50	2.14	2.51	2.49	1.64	2.09
Al_2O_3	15.68	14.59	15.90	13.78	12.97	14.72	14.74	13.23	12.44	13.95
Fe_2O_3t	11.65	15.14	12.16	12.81	12.35	12.35	12.14	10.93	10.80	12.50
MnO	0.20	0.21	0.15	0.18	0.16	0.15	0.17	0.14	0.12	0.13
MgO	4.38	4.25	6.51	5.80	11.24	7.08	7.81	6.90	11.35	3.17
CaO	13.90	8.63	7.01	7.19	5.67	8.02	8.19	9.46	9.86	11.06
Na_2O	4.35	4.90	4.40	3.15	3.06	3.27	3.74	3.80	1.03	2.71
K_2O	0.15	0.03	0.14	0.15	0.12	0.20	0.86	0.61	1.40	2.30
P_2O_5	0.25	0.86	0.12	0.14	0.25	0.31	0.58	0.51	0.16	0.33
LOI (1200°C)	9.89	6.29	3.12	4.70	7.06	9.32	4.32	3.64	12.69	11.01
Total	99.59	100.52	100.26	100.58	100.56	100.50	100.03	100.47	99.36	100.19
Trace elements (ppm)										
Ba	70	154	92	191	211	80	342	1446	72	303
Cr	32	19	273	246	534	39	248	275	408	49
*Cs	2.40	2.00	5.10	11.60	4.20	3.80	2.00	2.55	9.60	3.45
Cu	–	–	56	37	55	35	59	62	82	29
Ga	–	–	24	20	18	16	23	23	16	21
*Hf	4.7	8.0	3.3	2.5	3.2	5.0	5.0	5.8	3.2	4.2
Nb	23	51	6	7	18	22	43	44	23	29
Ni	41	16	89	81	381	64	135	107	244	46
Pb	–	–	6	8	20	11	12	13	11	15
Rb	4	2	6	10	6	15	19	12	30	58
*Sc	22.0	22.0	28.2	27.2	19.9	21.5	28.0	35.0	25.1	20.1
Sr	249	318	252	299	471	447	718	900	163	292
*Ta	1.91	3.68	0.40	0.40	1.36	1.70	2.88	2.83	1.37	1.83
*Th	1.86	3.72	0.90	0.50	1.70	2.06	3.90	3.97	1.90	1.96
*U	0.89	1.30	0.40	10.40	0.80	0.70	1.18	1.18	0.62	0.78
V	–	–	247	222	159	273	233	318	187	183
Y	20	34	16	26	17	21	24	25	24	33
Zn	–	–	99	109	104	99	106	100	98	101
Zr	152	264	87	95	129	155	205	215	125	171
Rare-earth elements (ppm)										
La	20.86	34.10	6.10	6.10	15.10	19.70	32.60	33.85	12.50	24.46
Ce	44.00	59.00	14.45	14.41	31.06	43.68	67.88	63.97	30.09	52.43
Pr	4.50	9.10	–	–	–	–	–	–	3.71	6.46
Nd	20.00	36.00	10.40	9.70	15.60	24.10	35.00	33.41	17.82	29.91
Sm	5.80	11.30	2.81	2.69	3.40	5.07	7.21	7.05	3.70	5.96
Eu	1.90	3.30	1.10	1.06	1.18	1.79	2.35	2.39	1.23	1.70
Gd	5.00	11.80	3.73	4.17	3.79	5.12	6.83	6.70	4.30	6.41
Dy	5.20	6.90	3.74	4.91	3.66	4.69	5.61	5.06	3.79	5.18
Ho	0.64	1.10	0.67	0.90	0.66	0.80	0.91	1.00	0.80	1.06
Er	1.90	3.70	1.86	2.57	1.82	2.10	2.27	2.30	2.23	2.95
Yb	2.00	2.90	1.53	2.21	1.60	1.72	1.71	1.66	1.82	2.28
Lu	0.34	0.35	0.22	0.32	0.24	0.24	0.24	0.26	0.25	0.30

Pentire alkali basalt lavas, samples 1 [Pentire Point, SW 932 808] and 2 [Kellan Head, SW 969 811]; anhydrous tholeiitic dolerites, samples 3 [Wadebridge, SX 014 715] and 4 [Park Head, SW 841 708]; anhydrous alkali dolerites, samples 5 [Stepper Point SW 915 784] and 6 [Rumps Point, SW 931 812]; hydrous alkali dolerites, samples 7 [St Saviours Point, SW 922 761] and 8 [Cataclews Point, SW 868 761]; Tintagel Volcanic Formation volcaniclastic rocks, samples 9 [Trebarwith Strand, SX 048 865] and 10 [Trebarwith Strand, SX 048 867].

 * Analyses by instrumental neutron activation analysis (Universities Research Reactor, Risley).

Rare-earth elements by inductively coupled plasma source spectrometry (RHBNC, Egham).

Major oxides and rest of trace elements by X-ray fluorescence spectrometry (Geology Department, University of Keele: analysts D Emley and M Aikin). Fe_2O_3t is total iron expressed as Fe_2O_3.

Pentire alkali basalt lava

This suite comprises the east-south-east-trending Upper Devonian lava outcrop and includes all inland exposures of lava to the east of the type locality at Pentire Point on the coast (Figure 9). The best exposures are along the coast from Pentire Point [SW 923 805] to Kellan Head [SW 970 812]. It is likely that greater variation (especially in chemical composition) is present throughout the mapped outcrop than described on the basis of the coastal sections alone.

The lava flows at Pentire Point are overturned and young southwards, a feature substantiated by pillow-lava packing features, such as gravity sag. Major joint sections in the steep cliffs show the morphology of the pillow lava as a series of draped, elongate and bulbous tubes with generally ovoid cross-sections. The coastal section of the Pentire peninsula indicates that there were at least two main eruptive events at different stratigraphical levels, and also a number of discrete vents feeding at least three pillow-lava mounds, separated by on-lapping mudstone.

These local features suggest that the whole lava outcrop may be composed of a series of small, fissure-fed, variably independent, pillow-lava domes, rather than a number of continuous pillow-lava sheets poured out over the whole sea floor.

The pillows in the coastal lava are typically small in cross-section (30 to 60 cm in diameter, rarely over 1 m) and are characterised by being highly vesicular (Plate 3b). Most show a series of concentric vesicular zones, some with a large central vacuole, that implies the passing of a series of lava pulses through the tube and its partial draining. The external vesicles are small (1 to 2 mm), and filled with secondary chlorite; towards the centre of the pillows the vesicles are generally larger (up to 5 mm) and filled with carbonate. Small black sediment fragments are trapped within the pillow tubes, and are similar to the cherty mudstone filling the interpillow spaces. The original glassy margins of the tubes are rarely preserved, having being spalled off during initial cooling, although rims are now represented by fine-grained dark chlorite. At some localities pillow breccia is associated with the lava, for exam-

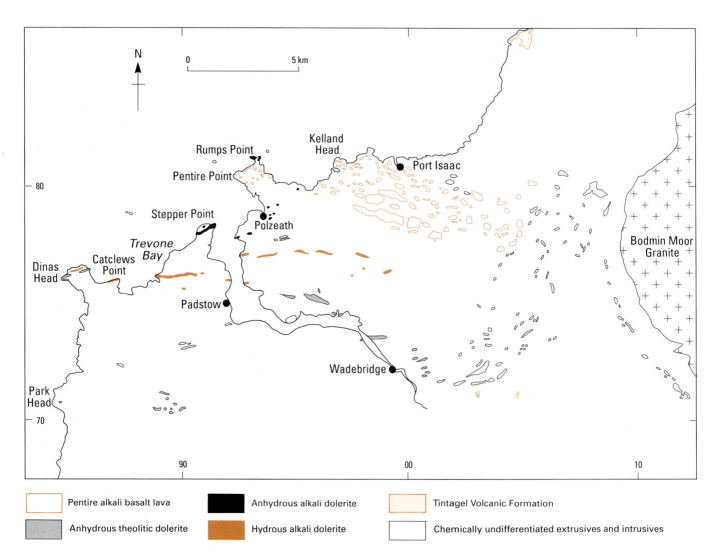

Figure 9   Sketch map showing the distribution of volcanic suites in the district.

ple, towards the base of Pentire cliff [SW 932 808]; this demonstrates the partial fragmentation of pillows along both radial and concentric cooling joints.

Petrographically the pillow lava is highly altered, but it retains textural features indicative of rapid quenching, typical of submarine lava accumulations. The lavas are moderately plagioclase-phyric alkali basalt with two size generations (<2 mm and >5 mm) of plagioclase (now albite, An_{2-8}) phenocrysts, both of which can occur as single crystals or glomerocrystic clusters. Some of the largest phenocrysts are seen at Kellan Head [SW 970 811] and exhibit crystal margin embayments due to resorption by the lava. Small plagioclase laths and microlites with forked terminations, or long curved serrated forms dominate the matrix, which is now composed of chlorite, variably oxidised iron-rich opaque minerals, rutile, sphene and carbonate. All original glass and mafic minerals have been totally replaced by secondary phases. Around some vesicles there are dark zones of fine-grained chlorite, without plagioclase microlites that probably represent the replacement of original quenched glass produced by rapid heat loss on endothermic vesiculation. Some of the carbonate infilling the vesicles is cloudy, with small inclusions. Electron microprobe examination shows that both the carbonate and the area adjacent to the vesicle are uranium rich (Williams and Floyd, 1981).

Geochemically the Pentire lava flows are a differentiated sequence of alkali basalt with variable but low abundances of compatible elements (Cr ranges from about 1 to 100 ppm), high concentrations of incompatible elements (Zr ranges from 100 to 450 ppm) and fractionated rare-earth-element (REE) patterns with $(La/Yb)_N$ about 8 to 9 (Figure 10). Olivine, pyroxene and plagioclase fractionation are necessary to explain the chemical variation exhibited by the lava, although plagioclase is the only observable phase in the exposed sequence. Two other chemical features are significant. Firstly, lava from Kellan Head is highly fractionated and comagmatic with most of the variably evolved lava exposed in the Pentire headland. Secondly, two chemical groups of lava with different Zr/Y ratios and incompatible element abundances appear to be present: those from Lundy Bay [SW 959 799] and the east side of Pentire Head [SW 936 808] with Zr/Y of about 10 and the remainder from Pentire peninsula and Kellan Head, with Zr/Y of about 7 (Figure 10). The overall chemical features displayed by the two groups reflects generation via different degrees of partial melting of a homogeneous source.

Anhydrous tholeiitic dolerite

This suite comprises various sill-like dolerite bodies emplaced in the Trevose Slate Formation and includes outcrops from Park Head [SW 840 708], Dinas Head [SW 847 862] to Merope Rocks [SW 860 766] in the west, and possibly scattered outcrops to the north-east and north-west of Wadebridge (Figure 9). Intrusive relationships are best exposed in the vicinity of Dinas Head, and indicate that the dolerite here has an irregular form with various separate offshoots penetrating the sedimentary

rocks at different levels. In the quarry above Dinas Head, the high-level emplacement of the dolerite into wet, partly consolidated sediments can be demonstrated. The upper contact is a vesicular basalt with a pillowy structure and flow-aligned ovoid vesicles now filled with dark chlorite. Other contacts with the sedimentary rocks are also indicative of reaction with wet sediments, such as very irregular or cuspate junctions and soft-sediment deformation structures. These features indicate that this suite has a similar age to the enclosing Trevose Slate Formation, late Middle Devonian or around the Middle–Upper Devonian boundary, and includes some of the earliest intrusions in the region.

This suite is distinctive in having a tholeiitic composition, with an anhydrous primary assemblage and sub-ophitic texture. Magmatic phases are clinopyroxene, plagioclase, ilmenite and accessory apatite, variably replaced by a low-grade metamorphic assemblage. The clinopyroxene is a colourless to pale pink augite typical of tholeiitic compositions with low concentrations of Ti, Al (Al/Ti formula unit ratios range from 2 to 4), Cr and Ni (Rice-Birchall, 1991). Some individual crystals are strongly zoned with cores rich in Mg+Ti+Al and margins rich in Fe. Their overall compositional range (from $Ca_{44} Mg_{41} Fe_{15}$ to $Ca_{42} Mg_{32} Fe_{26}$) is characteristic of pyroxenes from fractionated tholeiitic series (Figure 11). Many relict pyroxene crystals are fringed by secondary actinolite or completely pseudomorphed by chlorite and actinolite. Primary plagioclase is replaced by albite, together with variable amounts of carbonate, epidote, prehnite and sericite. Skeletal ilmenite may be oxidised (leucoxene), or replaced around margins by sphene; exsolved magnetite lamellae are locally present. The finer-grained facies of the sills are invariably entirely composed of secondary minerals.

The tholeiitic dolerite intrusions of this suite are chemically distinguished from the alkali intrusive rocks by exhibiting generally lower abundances of incompatible elements (Zr commonly <100 ppm), low tholeiitic Nb/Y ratios (<0.6; Winchester and Floyd, 1977), only mildly fractionated rare-earth-element patterns with $(La/Yb)_N$ about 1.9 and distinct, low Zr/Y ratios of about 4 (Figure 12). Although the range of FeO*/MgO, Ni and Cr contents is not very great, it does suggest that pyroxene and olivine fractionation were important in the petrogenesis of this series.

Anhydrous alkali dolerite

This is a relatively restricted intrusive suite occurring at Stepper Point [SW 915 785], Rumps Point [SW 933 812] (Plates 3c and 4) and an outcrop north of Pentireglaze to Lundy Bay [SW 857 799], together with small bodies associated with the Pentire lavas near Portquin. Like many of the other intrusive rocks this suite is dominated by sills, although a near-vertical dyke-like mass is seen at Stepper Point, with texturally variable chilled zones. Although this suite at first sight appears petrographically similar to the tholeiitic dolerites, it differs in bulk chemistry (being alkali) and in relict clinopyroxene composition, and is the only suite that can be geochemically related to the Pentire basalt lava flows (Rice-Birchall, 1991). Of all the dolerite intrusive rocks they are most often spatially associated with the lava, which indicates that this alkali dolerite suite is probably slightly younger than the tholeiitic dolerite suite.

The primary-phase mineralogy is typically clinopyroxene-plagioclase-ilmenite-apatite, although many bodies have been strongly deformed, internally sheared and re-

placed by extensive secondary assemblages (at Rumps Point, for example). Much of the internal variation in grain size and colour in the Stepper Point dyke is the effect of secondary alteration, although the marginal zones are chilled vesicular basalt with plagioclase phenocrysts and minute microlites that were originally preserved in a glassy matrix. The clinopyroxene is poorly fractionated augite with compositions similar to quench-textured

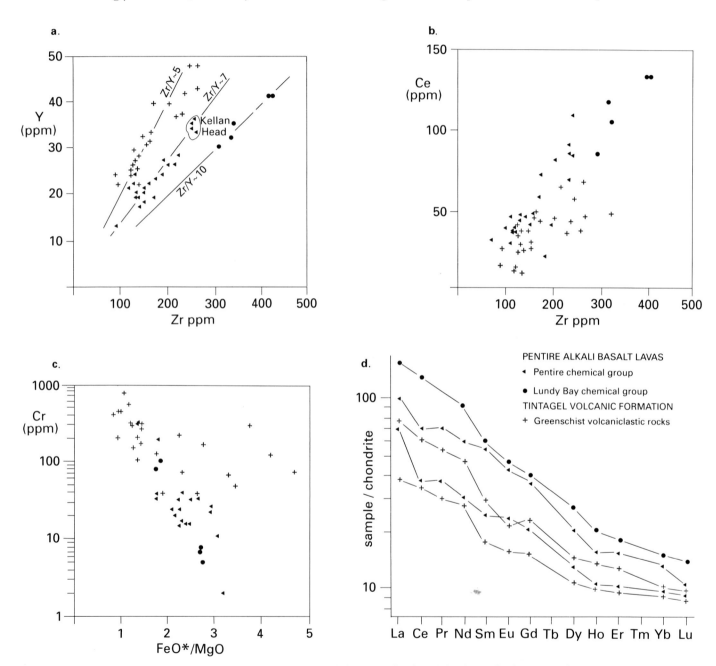

Figure 10   Selected geochemical features of alkali basalt lavas and volcaniclastic rocks from north Cornwall.

a. and b. show stable incompatible element variation and the development of different chemical suites with distinct Zr/Y and Ce/Zr ratios.   c. Illustration of the overall decrease in Cr with progressive differentiation (as crudely measured by the FeO*/MgO ratio, where FeO* is total iron expressed as FeO that reflects the operation of pyroxene fractionation.   d. Characteristically enriched (chondrite-normalised) rare-earth-element patterns for samples of the different alkali suites. Based on Floyd(1984), Rice-Birchall and Floyd (1988) and Rice-Birchall (1991).

a.

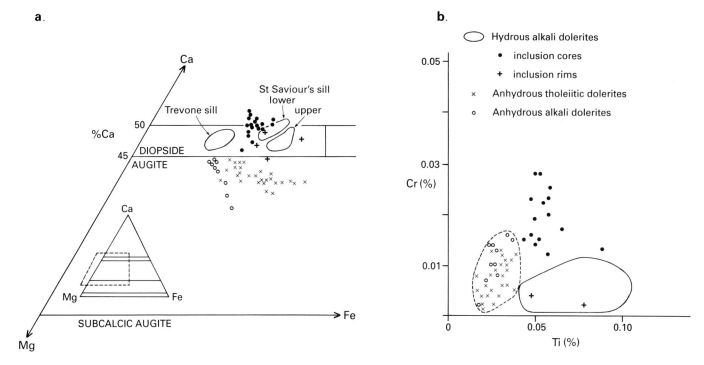

b.

Figure 11   Compositions of clinopyroxenes in dolerite suites from north Cornwall.

a. A portion of the pyroxene quadrilateral (see inset) shows compositional differences between pyroxenes from the hydrous dolerite sills (envelopes) and anhydrous dolerite sills (plotted points). The cores of pyroxenes from the high-pressure inclusions within the hydrous suite (dots) are clearly differentiated from the corresponding matrix pyroxenes. This feature is also illustrated in diagram b. by their generally higher Cr content b. Variation in two minor pyroxene components, Cr and Ti, the latter again distinguishes the hydrous and anhydrous suites. After Rice-Birchall (1991).

Plate 4   Rumps Point with The Mouls island in the background, viewed from the south-west [SW931 808].

A faulted flat-lying dolerite sill intrusive into Trevose Slate Formation forms the Point. Thin turbidite limestones near the sill yield deep subtidal conodonts of upper Givetian age. The deep gullies limiting the headlands to the south, mark late high-angle faults. In the foreground there are pillow lavas. (A15389)

species from tholeiitic basalts, having variable Ca but little change in Fe/Mg (Figure 11) and low Al and Ti contents. The bulk chemical composition, however, indicates that the dolerite is alkali, and suggests that the pyroxene composition has been influenced by early rapid growth rather than the bulk chemistry of the magma (cf. Floyd and Rowbotham, 1979). The alkali dolerite sills feature high Cr and Ni contents and generally lower incompatible element abundances than other alkalic suites in the region, and in general terms are less fractionated than the comagmatic Pentire lava. Zr/Y ratios and rare-earth-element fractionation patterns are distinct from the tholeiitic dolerite (Figure 12), although similar to the Pentire lava (Figure 10).

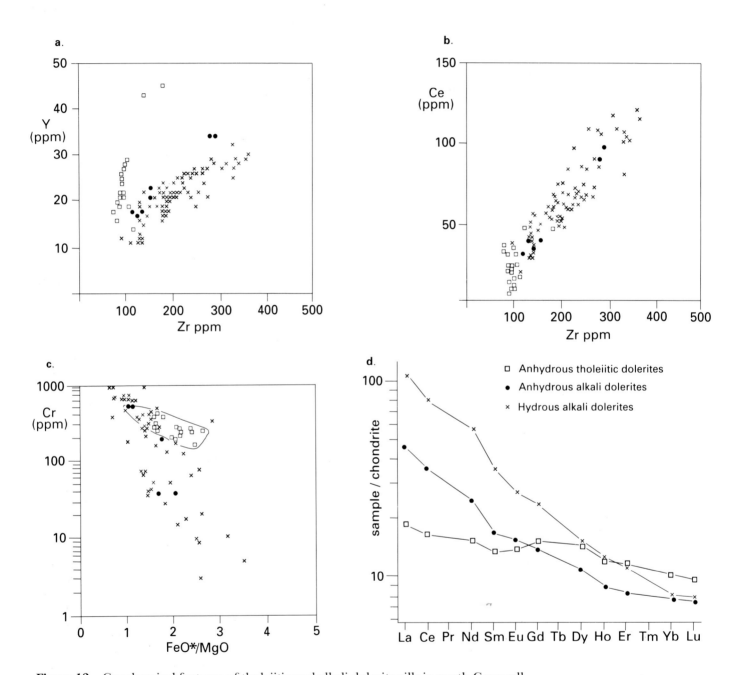

Figure 12   Geochemical features of tholeiitic and alkali dolerite sills in north Cornwall.
a. and b. show stable incompatible element variation and the distinction between the tholeiites, with generally low elemental abundances (except Y), and the enriched alkali dolerites. c. Illustrates a gross decrease in Cr with progressive differentiation (as crudely measured by the FeO*/MgO ratio, where FeO* is total iron expressed as FeO) that is a consequence of pyroxene fractionation in both tholeiites and alkali dolerites. d. Characteristically enriched (chondrite-normalised) rare-earth-element patterns for samples of the two alkali suites relative to the flatter pattern of a sample of the tholeiitic suite. Based on Floyd (1984) and Rice-Birchall (1991).

Hydrous alkali dolerite

This intrusive suite is mineralogically and chemically distinct from all the other doleritic rocks in exhibiting primary magmatic hydrous phases (amphibole and biotite), and generally higher abundances and more-variable ratios of incompatible elements. They were recognised in the earlier literature and generally refered to as 'proterobases' or sometimes 'minverites' after St Minver parish [SW 966 770] in this district (Reid et al., 1910; Dewey, 1914). They comprise two east–west-trending outcrops from Trevose Head [SW 850 766] through Cataclews Point [SW 873 762] and Trevone Bay [SW 890 763] to St Saviour's Point [SW 923 760] and Rock quarry [SW 931 758], and a smaller development to the north from Brae Hill [SW 927 772] eastwards (Figure 9). The major outcrop intrudes the upper part of the Trevose Slate Formation, and soft-sediment deformation features (at Rock quarry) indicate that some of the bodies are of a similar age, possibly early Frasnian, to the enclosing sedimentary rocks. Possible pillowy segments in the northern outcrop (coastal section at Brae Hill), suggest that these dolerites were emplaced during the deposition of the Famennian Polzeath Slate. A small outlier occurs at Com Head [SW 740 805] emplaced in Trevose Slate. All of the dolerite intrusions are sill- or sheet-like in attitude, although local lithological and chemical differences could suggest that the main southern outcrop (Trevose Head to St Saviours Point) may not be the disrupted remnants of a single body. At Rock quarry a thin (1 m wide) diagonal feeder dyke with irregular margins joins the base of the main body towards the top of the quarry face. The thicker dolerite intrusions vary considerably in grain size, from fine-grained chilled margins to gabbroic centres. Some have pegmatitic zones with large (10 to 15 mm long) black pyroxene prisms (for example at Trevone Bay [SW 890 762]); others exhibit coarse basal zones of crystal cumulates (for example at Cataclews Bay [SW 871 762] and Trevose Head [SW 854 768]).

The thickest member of this intrusive suite is St Saviour's sill (about 20 m thick), which, although it has been heavily dislocated during folding, is generally concordant with chilled margins, 0.75 m to 1 m thick, of vesicular basalt. It is a multiple intrusion with an internal chilled zone at about 12 m from the top, which divides it into upper and lower segments. Each segment shows a crudely similar change in lithology upwards from a lower olivine cumulate (>10 per cent olivine), to an olivine dolerite (1–3 per cent olivine), to the dominant clinopyroxene dolerite (low in olivine and amphibole), and finally to a leucocratic amphibole dolerite (amphibole>pyroxene) (Rice-Birchall, 1991). The olivine-rich layer of the upper segment is about 3 m above the base, and probably reflects flow concentration during the subsequent emplacement of the upper segment relative to the earlier lower segment.

The sill contains small inclusions of baked laminated sediment which are flow-orientated, parallel to the margins, and show mineralogical banding (mm to cm scale) of mafic-rich (including large brown biotites) and felsic-rich layers. There is a general coarsening upwards in grain size, together with a change to a more feldspathic lithology.

Petrographically the hydrous alkali dolerite intrusions are variably altered and replaced by prehnite-pumpellyite-facies assemblages, although the primary magmatic assemblage is composed of olivine, strongly pleochroic colourless to pink clinopyroxene, plagioclase, brown kaersutite amphibole, dark brown biotite, ilmenite and cored apatite needles (Reid et al., 1910; Dewey, 1914; Floyd and Rowbotham, 1982; Rice-Birchall, 1991). The characteristic feature of the primary assemblage is the highly Ti-rich nature of all the mafic and opaque phases. The clinopyroxene is titaniferous diopside with variable Mg/Fe ratios that reflect bulk compositional differences between different sills (Figure 11). This feature is also exhibited by the clinopyroxene from the upper, more chemically evolved segment of the St Saviour's sill which is more fractionated than the earlier lower segment (Figure 11). In terms of petrogenesis, textural and replacement relationships indicate near-equilibrium crystallisation under increasingly hydrous conditions and falling temperatures, with the successive development of olivine, clinopyroxene, (with ilmenite and apatite), amphibole and finally biotite (Floyd and Rowbotham, 1982).

An unusual feature displayed by the Cataclews sill is the presence of small, entrained, granular-textured, olivine-clinopyroxene (wehrlite) xenoliths (Rice-Birchall, 1991). The cores of the clinopyroxene crystals are Cr-rich calcic diopside, whereas the rims have a composition similar to the enclosing matrix clinopyroxene (Figure 11). The mineralogy and pyroxene chemistry of the exotic inclusions suggests they are the product of high-pressure fractionation prior to high-level crustal ponding and emplacement, when the rims subsequently re-equilibrated with the enclosing host magma.

Geochemical data for this suite include a few major oxide analyses in the early literature (e.g. Reid et al., 1910) and modern trace-element data comparing the hydrous and anhydrous dolerites (Rice-Birchall, 1991). This alkali suite is characterised and distinguished from the other dolerite instrusive rocks by generally high incompatible element abundances, greater variability of Nb/Y (c.1.5 to 3.0) and Zr/Y (c.8 to 10) ratios and well-fractionated rare-earth-element patterns with $(La/Yb)_N$ c.14 to 15 (Figure 12). The geochemistry is largely governed by olivine fractionation, followed by clinopyroxene+amphibole fractionation. In the more evolved (and freshest) samples especially high values of the large-ion-lithophile elements (e.g. K, Rb, Ba) reflect the abundance of amphibole and biotite. Olivine-clinopyroxene cumulates occur at a number of localities (Trevose Head [SW 854 768], Cataclews Bay [SW 871 762], St Saviour's [SW 923 760]), and are chemically distinguished by high Cr (>600 ppm) and Ni (>350 ppm) contents.

The spread of data and the variation of the Zr/Y ratio (Figure 13) suggest that the hydrous alkali dolerite intrusions may represent two chemical suites, comprising (a) Trevose Head, Trevone and Cataclews Bay (with

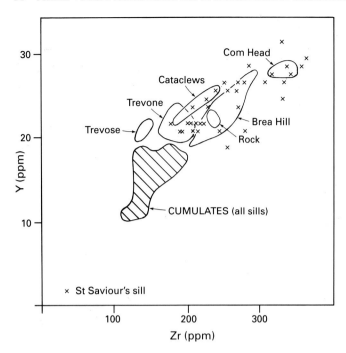

Figure 13 Distribution of Zr and Y of the hydrous alkali dolerite suite (named envelopes) and the St Saviour's Sill (crosses). Cumulates from all sills plot in the hatched envelope. After Rice-Birchall (1991).

Zr/Y c.8), and (b) Rock, Brae Hill and Com Head to the east of the Camel estuary (with Zr/Y c.10). Each chemical group also becomes more evolved or fractionated, with progressively increasing Zr and Y, from the west to the east. The cumulate compositions also reflect this apparent chemical trend, with the most primitive and highest Cr+Ni contents found in the west at Trevose Head. It is tentatively suggested that these features might indicate the eastwards propagation and injection of two separate fractionating alkali magmas, produced by slightly different degrees of partial melting. It should also be noted, however, that the full range of compositions (excluding cumulates) is almost matched by the internal variation of the highly differentiated St Saviour's banded sill.

LOWER CARBONIFEROUS

Tintagel Volcanic Formation

The Tintagel Volcanic Formation has undergone polyphase deformation and metamorphism to a series of banded, heterogeneous greenschists (Robinson and Read, 1981; Selwood and Thomas, 1986a). The overall degradation and alteration of the volcanic rocks makes it difficult to determine their original mode(s) of formation, although it is generally considered that they represent a sequence of volcaniclastic sediments with minor extrusive rocks (Freshney et al., 1972).

Although considerable textural variation has been induced by subsequent deformation of the Tintagel Vol-

canic Formation, those trace elements that are generally considered to be immobile during low-grade alteration indicate that the metavolcanic rocks represent a single, fractionated, comagmatic sequence of alkali basalt composition (Robinson and Sexton, 1987; Rice-Birchall and Floyd, 1988). Compared with the Devonian Pentire alkali lava flows, the Tintagel Volcanic Formation is more primitive in overall composition, with higher Cr and Ni coupled with lower abundances of incompatible element (Figure 10). Absolute rare-earth-element abundances are also low, although moderately fractionated patterns are common, with $(La/Yb)_N$ c.5–8 (Figure 10). A small Eu anomaly (Eu/Eu* c.0.8) in the most evolved volcanic rocks reflects the importance of late plagioclase fractionation. Although plagioclase is the only phenocryst phase still observed in the Tintagel Volcanic Formation, systematic element variation suggests that olivine and pyroxene were also important in the early stages (Rice-Birchall and Floyd, 1988). Zr/Y ratios (c.5) are relatively uniform (Figure 11), and are the lowest of all alkali suites (lava and dolerite) in the region. This feature, together with the chemical characteristics of the most primitive members, suggests that the parental melts which gave rise to the Tintagel Volcanic Formation were produced by higher degrees of partial melting than the other alkali suites in southwestern England.

REGIONAL SETTING AND TECTONIC ENVIRONMENT

Throughout the region all the volcanic rocks are associated with Middle to Upper Devonian and Lower Carbonferous basinal sediments, which are divided into four thrust-bound stratigraphical sequences—Padstow, Bounds Cliff, Liskeard and Tintagel successions (Figure 4). The basin is interpreted as a half-graben structure, with major fractures at the northern shelf-basin margin providing channelways for uprising basic melts (Selwood and Thomas, 1986b). From this model the linear belt of volcanics from the Padstow area eastwards are interpreted as the volcanic expression of progressive basin development marking the position of a major lineament.

A number of minor changes in chemistry and eruptive style can be observed in the volcanic rocks, through the Devonian and Lower Carboniferous. Early volcanism is represented by the presence of aquagene basic tuff (of unknown magmatic affinity), followed in the late Middle Devonian by the intrusion of high-level tholeiitic sills (anhydrous tholeiitic dolerite suite). Tholeiitic magmatism was apparently short-lived, since from Frasnian times onwards basic magmas are exclusively alkaline and were probably derived from a different mantle source (more chemically enriched) by generally lower degrees of melting. Frasnian and Famennian times are marked by extensive alkalic pillow lava, and minor associated sills and dykes (Pentire alkali basalt lava suite, and anhydrous alkali dolerite suite respectively), together with a chemically distinctive series of alkali sills bearing primary hydrous phases (hydrous alkali dolerite suite). The Lower Carboniferous also produced alkali magma (Tintagel Volcanic Formation) with a chemical composition similar to that of the Upper Devonian alkali lava. If the Tintagel Volcanic Formation is essentially a volcaniclastic sequence a change in eruptive style is indicated, with the development of possibly explosive submarine volcanism, relative to the quieter effusion of pillow lavas and high-level intrusives.

In comparison with the basic volcanic rocks of south and west Cornwall, those of this district exhibit the following distinctive geochemical features: (a) they are predominantly alkali basalts enriched in incompatible elements (Floyd, 1982), (b) they have distinctive incompatible element ratios, such that they form a separate magmatic province (Floyd et al., 1983; Floyd, 1984; Rice-Birchall and Floyd, 1988), and (c) they have features typical of within-plate basalts (Floyd, 1982, Rice-Birchall, 1991). Figure 14 shows the distinction between the alkali basalts of north Cornwall and the two predominantly tholeiitic provinces of south and west Corn-

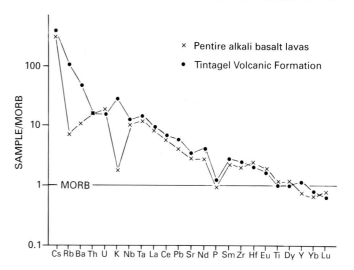

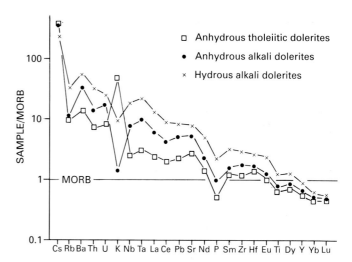

Figure 15 Multi-element patterns normalised to mid-ocean ridge basalt (MORB) for representative samples from basic volcanic suites in north Cornwall. MORB normalisation factors from Sun and McDonough (1989). Each pattern shows a gross progressive enrichment with increasing element incompatibility (Lu to Cs) relative to MORB and emphasises the within-plate (non-MORB) character of volcanic rocks from north Cornwall. Based on Rice-Birchall and Floyd (1988) and Rice-Birchall (1991).

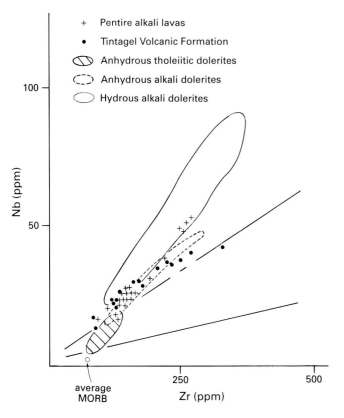

Figure 14 Distribution of Nb and Zr in basic volcanic rocks from north Cornwall compared with the other magmatic provinces of south-west England.

The relative distribution allows the identification of three major chemomagmatic provinces (field boundaries from Rice-Birchall and Floyd, 1988), with north Cornwall and Devon being characterised by the alkali lavas and intrusives described in this chapter. The associated tholeiitic intrusives, however, have chemical affinities with the south and west Cornish province and were probably derived from a different mantle source to the alkali group. Based on Rice-Birchall and Floyd (1988) and Rice-Birchall (1991).

wall, with which the anhydrous tholeiitic dolerite suite has features in common. The chemical discrimination of the eruptive setting as a within-plate tectonic environment supports current models of ensialic basin development for the Upper Devonian (e.g. Selwood and Thomas, 1986b), and underlines the absence of true oceanic crust in this region. As seen in Figure 15, both tholeiitic and alkali basalt/dolerite, irrespective of their mode of emplacement, exhibit similar incompatible element enrichment, characteristic of oceanic and continental within-plate basic rocks and distinct from mid-ocean ridge basalts.

METAMORPHISM

The metamorphic effects caused by or superimposed on the volcanic suites are widely separated in time, and comprise three events: (a) localised contact metamorphism of adjacent sediments by intrusive basic bodies during emplacement (Upper Devonian), (b) regional metamorphism of all volcanic rocks (early Upper Carboniferous, for details see Chapter six) and (c) contact metamorphism of some volcanic rocks within the thermal aureole of the Bodmin Moor granite (late Upper Carboniferous; for details see Chapter four).

Contact metamorphism

Two different contact effects can be recognised in the adjacent argillites, which reflect whether the sediments were relatively dry and consolidated, or wet and poorly consolidated at the time of intrusion. In the first case, contact effects are limited to progressive induration (within a metre or so), loss of colour (usually to pale grey), presence of thermal spotting and an increase in the prominance of small-scale features in the sediments (e.g. St Saviour's Point). The contact zone at Stepper Point is about 3 m wide and contains discrete beds with chlorite spots (2 to 3 mm in diameter) aligned in the direction of the S_1 cleavage. In the second case, the thermal effects are usually much more widespread, and are invariably accompanied by extensive Na-Si metasomatism (adinolisation). Soft-sediment deformation structures and irregular or cuspate margins between the intrusive and sediments on the small scale are common, and suggest that adinolisation may be a function of the emplacement level of the magma body. High-level intrusions penetrating soft, wet sediments, may provide heat that mobilises sea water Na trapped in pore spaces, which then becomes available for sediment autometasomatism. Generally, narrow adinoles are associated with a number of sills in the area, such as Cataclews, although more extensive developments are to be found. An adinole, about 40 m thick, is exposed on the eastern margin of a small intrusion at Tregildren's quarry [SX 035 793]. It contains pale albite-sericite-quartz clots set in a fine-grained foliated chlorite-sericite matrix. Of all the contact phenomena, however, the extended and variably metasomatised region related to the irregular Dinas Head intrusion is a classic example of contact metasomatism associated with dolerites (Fox, 1895b; Dewey, 1915; Agrell, 1939, 1941). At a distance from the contact, the initial effects are the presence of chlorite or incipient cordierite spots with andalusite prisms. Closer to the intrusion are partly metasomatised argillites (spilosites) with quartz-chlorite and albite-rich clots, which give way to completely metasomatised argillites (adinoles), composed almost entirely of a microgranular quartz-albite mosaic, adjacent to the contact (Agrell, 1939). At Dinas Head the above features are not always systematically observed as the contact is approached, because the bulk composition and grain-size of the sediment have influenced the degree of metasomatic replacement (Dewey, 1915) in such a way that spilosites and adinoles are often interbedded on the small scale.

FOUR

Granite and associated metamorphism

The Bodmin Moor Granite has an outcrop area of 190 km² of which about 70 per cent lies within the district. The outcrop is made up of coarse-grained biotite granite with several small areas of microgranite (fine-grained granite). There are pods and patches of leucogranite and xenolithic inclusions in the coarse-grained granite, and both varieties are cut by leucogranite and aplitic sheets of varying thickness and inclination.

The landscape of the Bodmin Moor Granite outcrop is typically open, undulating moorland, providing poor pasture for cattle, with cultivated areas along many of the valleys. The hills are generally crowned by tors (Plate 5a) and many hillsides have small, craggy outcrops, slabs, of granite in situ and boulders. Small, abandoned quarries and slabs worked in the open for gateposts, kerbs and the like are common.

The biotite-bearing granites of Devon and Cornwall have been classified into coarse-, medium- and fine-grained varieties, each group being further divided on the abundance of megacrysts, as shown in Table 2 (Hawkes and Dangerfield, 1978; Dangerfield and Hawkes, 1981; Goode et al., 1987). A modified version of this classification is adopted for descriptive purposes in the memoir. During the present survey, only granite and microgranite have been mapped in the Bodmin pluton. At some localities, small exposures and boulders show veins of fine-grained granite penetrating coarse-grained granite in an irregular manner, which suggests that the fine-grained granite is the younger of the two main granite types. Other granite types can, however, be recognised in individual outcrops and groups of outcrops. The boundaries between these varieties are commonly gradational and could not be mapped in detail. The distribution of granite types is shown in Figure 16.

The Devonian sediments surrounding the granite were regionally metamorphosed to greenschist or sub-greenschist facies before the intrusion of the granite. In the vicinity of the intrusion, thermal metamorphism has increased the metamorphic grade to upper greenschist facies and the lower level of the albite-epidote-hornfels facies. The thermal metamorphic aureole ranges in width from about 1.5 km round the south and west of the intrusion to about 2.5 km to the north where the dip of the contact is less steep.

REGIONAL SETTING

The Bodmin Moor Granite is one of six major and several minor exposures of granite plutons and satellite bodies comprising a batholith which underlies the whole of the south-western peninsula, from a few kilometres south-west of Exeter to beyond the Isles of Scilly, a total distance of nearly 250 km.

This major body, the chief British plutonic manifestation of igneous activity in the Rhenohercynian Zone, was emplaced at the end of the Variscan Orogeny. In other zones of the Variscan of the European continental mainland, equivalent granitoids occur in two series of which the older is dated 340–300 Ma and the younger 300–280 Ma. The older continental series tends to be associated with high-grade metamorphism and a metamorphic basement. The Cornubian granites are in the younger series associated with low-grade metamorphism. There is no exposure of the batholithic body itself which might lead to direct evidence of its origin.

The dimensions and shape of the batholith have been determined by a combination of seismic and gravity data. These show that, as far west as the Isles of Scilly, it is a tabular body with a steeper contact along its southern side than along the north, and with an unexposed ridge extending northwards from the Carnmenellis pluton to the St Agnes area. The width increases downwards to 40 to 60 km, and it has been suggested that the thickness increases eastwards from about 10 km in the vicinity of the Isles of Scilly to about twice as much immediately east of Dartmoor. The batholith is sited entirely within the crust, the Mohorovičić Discontinuity being placed at a depth of 27 to 30 km (Bott et al., 1958; Bott et al., 1970; Holder and Bott, 1971; Tombs, 1977; Brooks et al., 1983). It is separated into segments by north-west–south-east faulting, some of which was of pre-granite origin and was reactivated in the Tertiary (Dearman, 1963). There is a major, sediment-filled trough, probably following one of these lineaments, crossing the batholith between the St Austell and Carnmenellis outcrops (Tombs, 1977; Willis-Richards and Jackson, 1989). With that exception, the top of the batholith between the present main outcrops is 2 km or less below the present surface.

The true thickness of the batholith is difficult to determine since all calculations from geophysical observations depend on assuming constants for the rocks and their behaviour and, in the absence of direct knowledge of the rocks at depth, these assumptions cannot be verified. The contrast between the properties of granite and those of the surrounding crustal metasediments deceases as depth increases. Three seismic reflector horizons in the peninsula are described by Brooks et al. (1984). The deepest of these at 27 to 30 km, conforms well with the earlier estimates of the base of the crust. The shallowest reflector, at 7 to 8 km, is within the major granite plutons and has been interpreted as being due to the top of the zone to which stoped xenoliths have sunk (Bromley and Holl, 1986). The interme-

Plate 5a Logan Rock on Roughtor, viewed from the south-east [SX 1455 8078].

Weathering along joints of the upstanding granite has produced loose blocks that are rocking-stones (loganstones) on the top of the tor. (A15402)

Plate 5b Bodmin Moor Granite at De Lank Quarry [SW101 755]. The quarry face in De Lank Quarry, showing the method of rock sectioning using thermal lances and extrusion. At the bottom of the photograph is the quarry floor. (A15398)

diate reflector at a depth of 10 to 15 km is very persistent. It dips southwards and is present to the north of the batholith. This reflector is interpreted as a major thrust zone (Brooks et al., 1984). This interpretation agrees well with the general tectonic style of the area and provides a convincing floor to the granite, particularly as it seems to mark a significant change in seismic velocity. Although Shackleton et al. (1982) have suggested that granite magma travelled north along such a thrust zone, it seems possible that this horizon defines a level at which a large slab of crust, including granite, has been translated from its original site farther south.

The Bodmin Moor Granite was included in a study by Bott et al. (1970), who concluded that its density increased on its northern flank.

Age and origin

Radiometric determinations of the ages of the exposed Cornubian granites all indicate that emplacement took place less than 300 Ma ago. Data from Halliday (1980), Jackson et al. (1982) and Darbyshire and Shepherd (1985, 1987), using mainly K/Ar and Rb/Sr techniques, show two periods of intrusion. The earlier, approximately 295 Ma to 280 Ma ago, included all the major granite

Table 2 Main textural types of Cornubian granite (after Dangerfield and Hawkes, 1981).

Textural type	Mean matrix grain size	Abundance	Megacrysts size
Coarse	Larger than 2 mm	Abundantly megacrystic — more than 10%	Megacrystic — larger than 20 mm
		Moderately megacrystic — 5 to 9%	
		Poorly megacrystic — less than 5%	Small megacryst variant — smaller than 20 mm
Medium	1 to 2 mm	With few megacrysts	
		with very rare megacrysts	
Fine	Smaller than 1 mm	Megacryst-rich — more than 10%	
		Megacryst-poor — less than 10%	

outcrops, together with the smaller bodies of Carn Marth (near Carnmenellis), Hingston Down and Kit Hill (between Bodmin Moor and Dartmoor) and Hemerdon Ball (south of Dartmoor). The second period, between 280 and 270 Ma ago, coincided with the onset of the main period of mineralisation, and includes the Li-mica granite at Tregonning, biotite granite reset by the intrusion of Li-mica granite at Castle-an-Dinas (north of St Austell), some elvan dykes and mineral veins.

Stratigraphical evidence for the age of the Bodmin Moor Granite shows only that it is younger than the late Devonian Tredorn Slate. However, as part of an Rb/Sr study of the radiometric ages of granite magmatism in south-west England, Darbyshire and Shepherd (1985) included nine specimens from Bodmin Moor, collected from West Carbilly Quarry [SX 125 754], Hantergantic Quarry [SX 103 757], De Lank Quarry [SX 101 755], Cheesewring Quarry [SX 259 724], Siblyback dam wall [SX 229 703] Bearah Tor Quarry [SX 259 745], a quarry on the west side of Caradon Hill [SX 269 705] and Craddock Moor Quarry [SX 249 724 and SX 249 and 723]. These samples gave a whole-rock isochron age of 275 ± 9 Ma and an initial $^{87}Sr/^{86}Sr$ ratio of 0.7166 ± 0.0021, which the authors regarded as unsatisfactory because of the large errors. Analysis of potassium feldspar, plagioclase, biotite and muscovite from one of the specimens from Caddock Moor Quarry [SX 249 723] gave an $^{87}Sr/^{86}Sr$ initial ratio of 0.7140 ± 0.0002 and an isochron age of 287 ± 2 Ma, which Darbyshire and Shepherd (1985) believed to be the best estimate of the age of emplacement, although they warned that it might have been reset.

More recently, Chesley et al. (1993) have determined a $^{207}Pb/^{235}U$ age of 291.4 ± 0.8 Ma for monazite in granite from a deep cutting [SX 162 742] adjacent to the A30 trunk road. The same specimen produced an $^{40}Ar/^{39}Ar$ muscovite plateau age of 287.1 ± 0.9 Ma and another, from De Lank Quarry [SX 101 755] an age of 288.1 ± 0.8 Ma. These determinations form part of a study of the cooling history of the Cornubian Batholith; in the case of the Bodmin Moor Granite a cooling rate of 95°Cmyr^{-1} is calculated, based on a closure temperature of 725°C for U-Pb in monazite and of 320°C for K-Ar in muscovite. The Ar-Ar ages given by Chesley et al. (1993) are in good agreement with the minerals isochron of Darbyshire and Shepherd (1985).

The chemical, normative and modal compositions of the exposed south-west England granites show them to be peraluminous and to have relatively high K/Na ratios, low Fe_2O_3/FeO ratios, high normative corundum, and high initial $^{87}Sr/^{86}Sr$ ratios. They contain muscovite, ilmenite and sometimes cordierite, but no magnetite, hornblende or sphene. These are all characteristic features of the 'S' Type granites described by Chappell and White (1974): that is, they indicate an origin from crustal metasedimentary rocks rather from the differentiation of a more basic igneous magma. This is consistent with the geophysical evidence, which has shown the batholith to be situated in a mid-crustal region, and has not shown the presence of a large body of basic rock which could be associated with differentiation. It should be noted, however, that the granite may have been moved from its source along the seismic reflection horizon at a depth of 10–15 km, which Brooks et al. (1984) interpreted as a major thrust.

Information about the source rocks themselves is sparse. The granites contain pelitic xenoliths, but these are mostly of low metamorphic grade and have been derived from high crustal levels. Of more help are sillimanite-bearing xenoliths (Ghosh, 1927; Jefferies, 1985, 1988) which suggest that pelitic rocks were present at depth, a suggestion supported by high ^{18}O and ammonium concentrations in the granite (Sheppard, 1977; Jackson et al., 1982; Hall, 1988). There are grounds, therefore, for suggesting that the magma originated by the partial melting of water-poor garnet granulite (Charoy, 1979; Floyd et al., 1983; Stone and Exley, 1986; Stone, 1988) or cordierite-sillimanite-spinel gneiss, crustal materials described by Charoy (1986) as resembling Brioverian basement.

The depth at which this partial melting took place presents another uncertainty, and estimates have to reconcile the contemporary thickness of the crust and geothermal gradient, the amount of subsequent erosion and the mineralogy of both granite and associated metamorphic rocks, bearing in mind probable shifts in solidus temperatures resulting from high volatile concentrations in the magma. Bromley (1989) has reviewed the arguments of a number of authors and has concluded that the crustal thickness was about 35 km, that the geothermal gradient was between 40°C and 45° C/km, that between 4 and 7 km, of crustal material has been eroded, and that partial melting occurred at a depth of 18–22 km. He also concluded that this removed the ne-

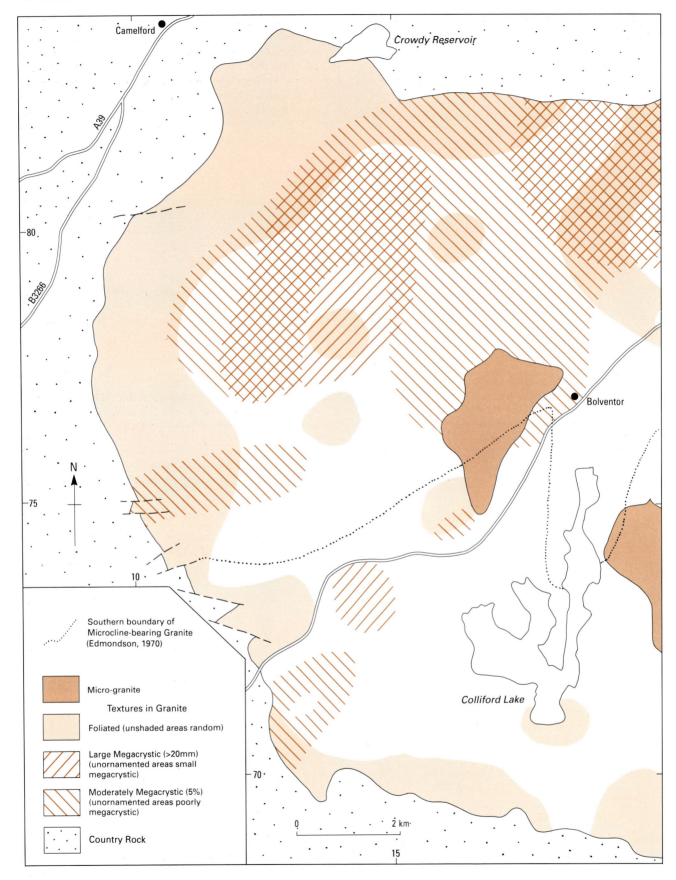

Figure 16 Distribution of granite rock types in the district (based on information from J Dangerfield and J B Hawkes).

cessity for a southerly source region and transport on southerly dipping thrust planes, suggesting that the intermediate seismic reflector at 10 to 15 km marked the top of the zone of partial melting.

Whether the magma originated beneath the present batholith or elsewhere, two peculiarities of the granites need explanation: their high levels of metallic and volatile elements, and their high thermal output. All the granites of south-west England are relatively rich in As, Cu, Cs, Ga, Li, P, Pb, Rb, Sn, Ta, Th, U, W and Zn and in B, Cl and F. Many of these elements have probably been derived through the assimilation of sedimentary rock or the passage of water drawn from surrounding sediments. Others, for example Sn, Ta, U, W, Cl and F are unlikely to have accumulated in this way, at least in the concentrations observed, but may derive from direct addition from a subcrustal source, or incorporation from an already enriched crust (Lister, 1984; Watson et al., 1984; Charoy, 1986).

The high heat productivity of the batholith results largely from its potassium content, and radioactive elements in the accessory minerals monazite, apatite, xenotime, zircon and uraninite: a major contribution being made by the last. Calculations by a number of authors, however, have tended to suggest that the present heat output could not be achieved by these minerals alone and that there must also be a subcrustal contribution (Simpson et al., 1979; Watson et al., 1984). Studies by Wheildon et al. (1981) and Lee et al. (1987) indicate this contribution to be 25 to 30 mW/m^2, a figure accepted by Willis-Richards and Jackson (1989) in their calculations and modelling of heat production and flow in the Carmenellis area of south Cornwall.

Most of the recent discussion of the origin and development of the batholith has been based on evidence derived from, and arguments restricted to, particular plutons; but conditions and processes were not uniform throughout its length. The Carnmenellis and Bodmin Moor plutons contain granites, which are similar in age, level of erosion and rare-earth-element patterns; they differ in all these respects from the other granites, and probably originated from different source rocks with different evolutionary histories (Darbyshire and Shepherd, 1987).

Evolution

The most important consequence of the high heat-production character of the batholith is its continued development after its initial formation. This may have resulted in the evolution of new types of rock, including Li-mica granite, fine-grained granite and elvan, and in extensive mineralisation, the ages of some of which have been measured at about 270 Ma.

The Li-mica granites of Tregonning, St Austell and Dartmoor belong to this period (Bristow and Exley, 1994), as do those elvans which have been dated. No fine-grained granite has yet been radiometrically dated, but it must be younger than the main granite which it intrudes. Willis-Richards and Jackson (1989) have calculated that if the batholith were blanketed by more than 2 km of covering

rock, cooling after initial emplacement would take 10 to 20 Ma. During this time a pool of residual magma above the solidus temperature would remain and, being enriched in volatile and metallic constituents, would be capable of producing a second magmatic and mineralising phase. Bromley (1989) has shown that the finer-grained granitic variants could have been derived by the dehydration of foundered xenoliths enveloped by the original magma during intrusion.

Younger mineralising events, including argillation, occurred in south-west England at intervals from about 250 to 100 Ma ago, with a further event at about 70 Ma (Jackson et al., 1982). These are attributed to convection systems of meteoric water, driven by heat from the batholith. There is evidence that this circulation still takes place.

GRANITE

The granite of Bodmin Moor is coarse grained and megacrystic with a grain size averaging 2 to 3 mm. It has a typically inequigranular, hypidiomorphic, granitic texture and is grey when fresh becoming buff when weathered. Megacrysts are chiefly euhedral to subhedral potassium feldspar and anhedral quartz. The former sometimes attains more than 20 mm in length and the latter 10 mm, but the abundance and size are variable. The megacrysts often have a random orientation but some are sufficiently well aligned to impart a degree of foliation to the rock.

The principal textural varieties (Table 2 and Figure 16) are as follows:

Coarse-grained biotite granite with small megacrysts (less than 20 mm)

The greater part of the outcrop is occupied by this variety. It is surrounded by and includes patches of the other types. The texture is typically granitic, and the rock is locally mineralised. It is poorly megacrystic with less than 5 per cent of megacrysts.

Moderately and abundantly megacrystic granite with small megacrysts

These varieties, respectively with 5 to 9 per cent and more than 10 per cent of megacrysts, occur in irregular patches. Examples occur around Metherin Downs [SX 1200 7450], near Rough Tor [SX 1450 8050], and Bray Down [SX 1850 8200].

Coarse-grained megacrystic biotite granite with megacrysts longer than 20 mm

This type occurs in patches within the small megacryst variant. Examples occur west of Temple [SX 1400 7350], in the vicinity of Casehill Downs [SX 1250 7850] and on Rough Tor and Bray Downs. Some is poorly megacrystic and some moderately or abundantly megacrystic.

Foliated granite

In appearance this is the most distinctive type in the district and represents a textural variation of the typical small megacrystic to poorly megacrystic variant. It occupies an arc of irregular width almost all the way around the granite margin (Figure 16), with isolated patches also being found around Colliford Down [SX 1800 7100], Brockabarrow Common [SX 1600 7500], Carkees Down [SX 1400 7650], Priest Hill [SX 1400 7800] and Brown Willy [SX 1600 8000].

The modal compositions of these main granite types are given in Table 3 and specimen localities in Figure 17. Although their mean compositions are similar, there are wide differences within each textural type.

Quartz is the most abundant mineral (Table 3), occurring both as megacrysts and in the matrix. Megacrystic grains are uncommon; they are invariably composite and in rounded aggregates up to about 10 mm across. In the matrix, it is in anhedral, irregular and interstitial forms. Most commonly it forms crystals or composite grains 1 to 3 mm across, although some may be much smaller, down to 0.2 mm, and some larger, up to 5 or 6 mm. The margins of individual crystals are commonly sutured, and the crystals themselves strained and contain trains of inclusions. Small areas of crushing occur in a few specimens taken from places near faults as, for example, on Cardinham Moor [SX 1339 7157], Buttern Hill [SX 1778 8166] and near the dam at Crowdy Reservoir [SX 1399 8330].

The megacrysts, as well as a substantial proportion of the matrix, are orthoclase microperthites. Megacrysts are subhedral, their apparently euhedral shape disguising the fact that their margins are generally irregular, partly enveloping adjacent crystals, and lack some crystal faces. Strings, patches and braids of albite, in places twinned, characterise the microperthite. More rarely the albite traverses the boundaries of the host crystals. Antiperthite occurs in a specimen from Codda Tor [SX 1756 7926]. Megacrysts enclose most other minerals, but most commonly quartz, plagioclase and mica. These inclusions are small and are, in some instances, arranged in zones parallel with the margins. They are mostly anhedral or sub-

hedral, and in the case of plagioclase may exhibit both twinning and compositional zoning. All the orthoclase shows simple twinning, although the twin planes may be 'stepped'. Reaction intergrowths occur with adjacent quartz to give graphic or granophyric textures, or with plagioclase to give myrmekitic textures.

Potassium feldspar in the matrix may be orthoclase or microcline, is subhedral to anhedral, irregular and often made up of aggregates of two or three crystals. As with the megacrysts, the margins commonly envelop contiguous minerals. Sizes are very variable; individual crystals range from about 0.1 to 6 mm, but are mostly 2 to 5 mm. Like the megacrysts, the larger matrix crystals often enclose quartz, plagioclase and mica.

South of an irregular line extending from near The Beacon [SX 1040 7400] via Durfold [SX 1170 7380] to Dairywell [SX 1820 7740], south to Gillhouse Downs [SX 1880 7390] and thence to the eastern boundary of the map, the potassium feldspar in the matrix is orthoclase with a 2V averaging $62.2°$. North of this line the potassium feldspars are microcline with 2V varying between $56.5°$ and $63.3°$ (Edmondson, 1970).

All the potassium feldspar in the granite has been altered, to some degree, to secondary white mica, in the form of small flakes or aggregates of flakes which follow cleavages and fractures, and some link with muscovite flakes outside the feldspar.

Plagioclase is always less abundant than potassium feldspar and occurs in more regular, euhedral to anhedral crystals, by far the greatest number being subhedral. The usual size is 1 to 3 mm, with some as small as 0.2 mm and some as large as 6 mm. Some grains consist of aggregates of several individual crystals. Multiple twinning is universal and follows the usual albite law; Carlsbad and pericline twinning are much less common. Strain is sometimes exhibited by the bending or wedging of twin lamellae.

All the plagioclase is normally zoned, from cores mostly of An_{25-30} to outer zones of approximately An_5; some have rims of clear albite, whereas the cores of the crystals are invariably altered in varying degrees to aggregates of mica and clay minerals, and this makes the determination of composition difficult.

Although these rocks are described as biotite granites, muscovite is always present, in amounts approaching 10 per cent. It forms flakes and plates up to about 4 mm in length, mostly between 1 and 2 mm. Individual flakes are commonly in aggregates, and muscovite and biotite are sometimes intergrown. Flakes from rocks near fault zones may be bent.

Biotite is present in anhedral plates and flakes with ragged ends, and ranges in size from about 0.3 mm to 3 mm. It is, in part, intergrown with muscovite and in places forms clusters of flakes. It is red-brown where fresh, but where mineralisation has occurred it is a brown or khaki-yellow colour. It is almost everywhere chloritised to some extent and, like muscovite, is bent in specimens from near faults. Biotite displays pleochroic haloes where it encloses radioactive minerals.

A fine-grained secondary white mica replaces potassium feldspar, muscovite and andalusite. It is found in

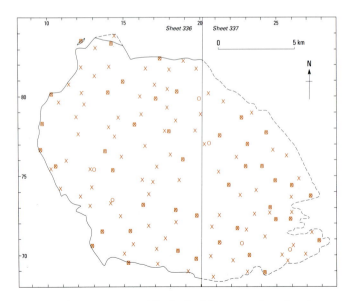

Figure 17   Outcrop of Bodmin Moor granite, showing locations of modally (X) and chemically (O) analysed specimens of coarse-grained granite.

Table 3   Modal analyses of Bodmin Moor granites.

	Ga	Gb	Gc	Gd	Ge	Gf
Number of samples	23	9	1	13	29	5
Quartz	33.5 (24.3–41.0)	35.2 (29.3–42.9)	32.3	32.0 (27.8–39.2)	31.9 (25.2–42.8)	33.7 (31.1–37.8)
Potassium feldspar	28.2 (21.4–37.8)	23.5 (14.8–29.5)	26.9	28.6 (19.1–33.8)	27.1 (21.4–33.4)	25.3 (18.4–29.6)
Plagioclase	16.2 (8.9–24.3)	18.2 (14.5–22.9)	20.6	17.1 (14.1–19.5)	18.1 (7.2–25.0)	23.4 (18.9–29.8)
Muscovite	8.3 (4.5–15.9)	8.5 (6.5–12.1)	5.3	7.4 (5.1–8.9)	8.1 (4.7–17.2)	9.0 (6.1–12.8)
Biotite	3.0 (0.5–6.8)	4.4 (1.3–6.6)	5.0	4.1 (1.7–7.2)	4.2 (1.1–7.3)	2.3 (0.3–3.6)
Secondary mica	0.9 (0.3–2.6)	0.3 (0.1–0.9)	1.3	0.6 (0.1–1.3)	0.8 (0.2–1.6)	0.5 (0.2–0.9)
Clay	8.3 (1.5–15.3)	8.0 (5.3–12.4)	7.0	8.9 (5.8–10.7)	8.8 (4.5–13.6)	4.4 (2.1–7.7)
Apatite	0.3 (0.1–0.5)	0.3 (0.1–0.5)	0.4	0.3 (0.1–0.4)	0.3 (0.1–0.5)	0.2 (0.1–0.3)
Tourmaline	1.2 (tr–3.0)	1.2 (0.3–2.4)	0.7	0.8 (0.4–2.0)	0.6 (0–2.0)	0.9 (0–2.7)
Andalusite	0.1 (0–0.6)	0.2 (0–0.5)	0.3	0.1 (0–0.5)	tr (0–0.3)	0.2 (0–1.1)
Opaque minerals	0.1 (0–0.2)	tr (0.1–0.1)	tr	tr (0–0.1)	0.1 (0–0.5)	n.f.
Zircon	tr (0–0.1)	0.1 (0–0.1)	0.1	0.1 (0–0.2)	0.1 (0–0.1)	tr (0–tr)
Fluorite	tr (0–tr)	tr (0–tr)	n.f.	n.f	tr (0–tr)	n.f.
	100.1	99.9	99.9	100.0	100.1	99.9
Potassium feldspar: plagioclase	1.20 (0.55–3.76)	0.91 (0.42–1.52)	1.00	1.11 (0.65–1.73)	1.02 (0.65–3.17)	0.92 (0.49–1.45)
IC	27 (20–37)	28 (23–33)	28	27 (22–35)	30 (23–41)	103 (62–170)

Original potassium feldspar calculated as observed potassium feldspar + $\frac{1}{2}$ 2y mica
Original plagioclase calculated as observed plagioclase + clay
Ga = Small megacryst variant, poorly megacrystic; Gb = Small megacryst variant, moderately and abundantly megacrystic;
Gc = Megacrystic, poorly megacrystic; Gd = Megacrystic, moderately and abundantly megacrystic; Ge = Foliated; Gf = Fine-grained; Ga – Ge all coarse-grained.
Compositional variation of samples within type is shown in brackets beneath mean values.
IC = Chayes IC number (Chayes, 1956).

very small flakes, never more than a few tenths of a millimetre in length, which may occur individually, for example in feldspar cleavages, or in aggregates, strings or veinlets.

Tourmaline is always present in the granite, although in variable amounts. It is generally anhedral but is also present in subhedral, irregular prisms, some of which, although discontinuous, may be recognised as parts of radiating 'suns'. Discrete anhedral grains are usually 0.3 to 1 mm across, but may form clusters up to 3 mm across; individual prisms may be up to 3 mm in length with basal sections 0.3 to 1 mm across. The predominant colour in thin section is yellow-brown, but irregular patches and zones of blue are not uncommon. Tourmaline replaces and is intergrown with all the major mineral phases, and is therefore thought to be of late formation.

A dense intergrowth of minute flakes, mostly of kaolinite, which may be iron-stained in places is often seen to replace plagioclase.

Apatite is the second most abundant accessory mineral after tourmaline and is ubiquitous. It is usually anhedral, rounded or irregular in shape, although some subhedral grains exhibit prism and basal sections. The former may be up to 0.4 mm long and the latter 0.2 to 0.3 mm across. Most grains are clear but some are cloudy, with minute dark inclusions.

Andalusite occurs sporadically in small amounts as anhedral, commonly irregular crystals, sometimes in groups where larger crystals have been separated along cleavage planes by alteration. Some of these are prism sections, others basal sections; they are mostly 0.1 to 0.3 mm in size, but some reach 0.5 mm. The mineral is pink in colour; it is commonly associated with muscovite, and altered to secondary mica around its margins and along cleavages.

Opaque minerals comprise minute grains of iron oxide and needles of ilmenite, produced in biotite as a result of alteration. They seldom occur elsewhere.

Zircon occurs in biotite in minute crystals surrounded by pleochroic haloes. Other radioactive minerals may also be present, but they cannot be recognised by ordinary optical methods.

Fluorite has been observed in trace amounts in specimens from a borehole [SX 1668 7468] on Sprey Moor and from Hawk's Tor [SX 1407 7544], Dinnever Hill [SX 1221 7950] and Codda Downs [SX 1690 7980], all areas close to mineralised zones. The mineral occurs as small purple grains in cleavages in mica and interstitially.

Strongly foliated granite with scarce megacrysts is present in some parts of the Bodmin Moor Granite, although it has not been mapped separately. Its mean modal composition is close to that of the other coarse varieties, although its overall grain size indicated by the Chayes IC number (Chayes, 1956) is slightly less coarse (Figure 18). The foliation is the result of deformation which has recrystallised the quartz and much of the feldspar, so that the latter is roughly aligned and enclosed in a mosaic of small, strained quartz grains with irregular, sutured margins. The degree of foliation is variable; in some areas, for example Leskernick Hill [SX 1830 8028], it is reduced to small areas of crushing. There are very few euhedral grains; most of the potassium feldspar is anhedral, often irregular in shape, and the proportion of anhedral and distorted plagioclase grains is greater than in the non-foliated rocks.

Muscovite and biotite flakes in the foliated granite are commonly bent, and biotite is frequently partially altered to chlorite, in places with an accompanying development of secondary potassium feldspar in lens-shaped crystals in the cleavages. A specimen from near Highsteps [SX 1116 8084] on the line of the major north-east–south-west fault in the north of the granite, approaches an augengneiss, with mylonitisation, recrystallisation and considerable chloritisation. Rocks from the west side of the Devil's Jump [SX 1009 8016] and 0.7 km to the southeast [SX 1054 7963] are also severely affected; their mineralogy is igneous but their fabric is essentially metamorphic. At the western margin of the pluton, the rock at Hamatethy [SX 0948 7829] is similar to that at Highsteps but the degree of crushing can be seen to diminish southwards through Row [SX 0935 7670], where it is similar to that near the Devil's Jump, to De Lank [SX 1004 7550] and to Pendrift Downs [SX 1066 7428]. At Corner Quoit [SX 1254 7082], there is little evidence of metamorphism, and the foliation, which is still faintly perceptible, is due to small amounts of movement, probably imposed on a magmatic orientation. The distinction between northern, microcline-bearing, and southern, orthoclase-bearing rocks applies to the foliated granite in the same way as to the non-foliated varieties.

Figure 18   Isopleth maps of features of coarse-grained granites.

a. Trend surface of Chayes' I.C. Number.
b. Trend surface of modal biotite.
c. Distribution of modal andalusite (raw data).
d. Trend surface of modal tourmaline.
e. Trend surface of Mafic Index.
f. Trend surface of ratio potassium feldspar/plagioclase.

Microgranite

Microgranite forms three areas of the Bodmin Moor Granite in the district. The most easterly extends from near Brown Gelly [SX 2000 7220], around the west of Dozmary Pool [SX 1880 7420] and to the north of Little Wood [SX 1970 7570]. The second area lies to the west of Bolventor, extending from Sunning Hill [SX 1660 7550] in the south to Tolborough Downs [SX 1700 7790] in the north, and from Dairywell [SX 1820 7740] in the east, to the streams west of Priddacombe [SX 1585 7635] in the west. In a shallow quarry near Priddacombe [SX 1645 7635], fine- and coarse-grained granite occur together in patches, which makes their true relationship difficult to determine.

The third area, not shown on the 1:50 000 geological map, is difficult to define. Loose material in a hollow between Roughground [SX 1045 7670], Penquite [SX 0980 7630], Row [SX 0960 7650] and St Breward [SX 0980 7725] was mapped by Reid et al. (1910) as a roughly circular area nearly 1 km in diameter. This survey suggests that it is much smaller than this, and consists of a complex of fine- and coarse-grained granites with veins and apophyses.

The microgranite is either non-megacrystic or very poorly megacrystic, but with a hypidiomorphic, inequigranular-seriate texture in which only plagioclase approaches a euhedral shape. The largest crystals are up to 2 mm across, but most range between 0.5 and 1 mm in diameter. A specimen from near Roughground [SX 1038 7640] in the western, fine-grained exposure shows some evidence of crushing.

Quartz is anhedral, irregular, both rounded and interstitial and often in aggregates or composite grains. Potassium feldspar is also anhedral and irregular in shape, in part interstitial, and in composite grains. It includes the other main minerals in some places and has graphic intergrowths with quartz at some margins. Most plagioclase is subhedral or anhedral and irregular, with rare euhedral crystals. It is cloudy and normally zoned. Both micas form typically ragged flakes, and biotite is usually chloritised to some extent. Apatite is rounded, yellow brown tourmaline forms irregular grains and prisms and andalusite is anhedral and pink.

Compositional variations

The coarse-grained granite is a two-mica adamellitic rock (Tables 3 and 4) which lies mostly in the syenogranite and alkali granite fields of Streckeisen's (1976) classifica-

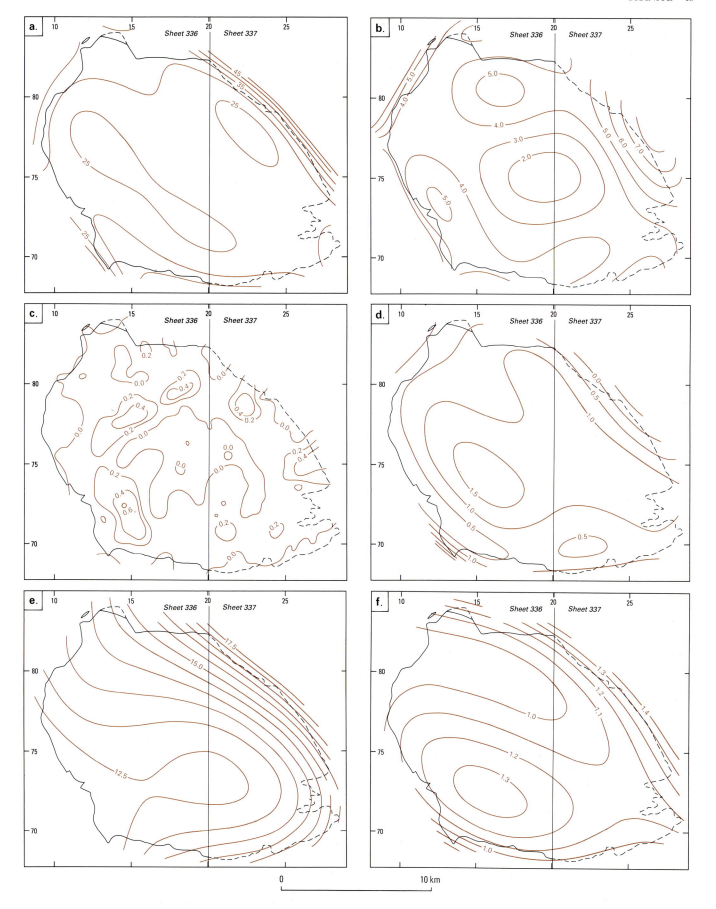

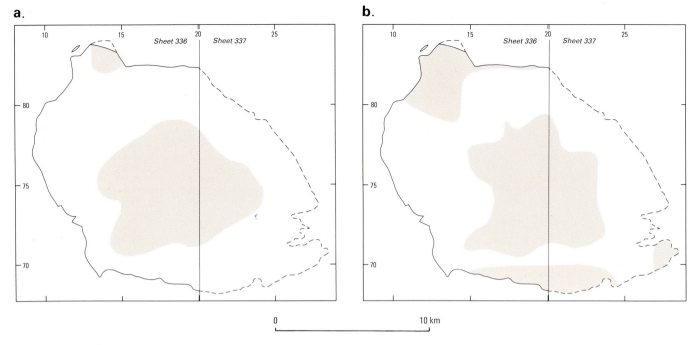

Figure 19   Maps of coarse-grained granite. Shaded areas showing approximate areas where rocks have   a. Fe_2O_3 + MgO less than 1.8%, and   b. normative anorthite less than 2%.

tion. The mean modal and chemical compositions* of samples from this granite are so similar that they seem to represent a chemically homogeneous magma. However, the ranges in composition are such that there is the possibility that the mean values conceal significant variations which are not distinguishable by simple inspection of the analyses, nor are they distinguishable in the field as separate intrusions.

Two statistical approaches have been used to investigate these variations, both dependent on relatively large numbers of analyses. A total of 122 modal analyses, each determined by point-counting of more than of 4000 points, and 52 chemical analyses, were made over the whole granite outcrop from sites not less than 500 m apart (Figure 17). The modal analyses (Table 4) were subjected to trend-surface analysis (cf. Davis, 1973) to distinguish regional changes. The major elements in the chemical analyses (Table 4) were grouped into clusters by Q-mode factor analysis to reveal their similarities, and some elements were plotted on binary diagrams to find other groupings.

The common feature of the analyses of modal compositions of the Bodmin Moor Granite (Figure 18) is the depiction of a roughly concentric pattern. Biotite (Figure 18b) and andalusite (Figure 18c) have their lowest values in the central region where contamination is least. Tourmaline (Figure 18d) has its highest concentration in the south-western part, in the general area of the intersection of two major fracture zones. It seems, however, to have had little effect on the shape of the trend surface of the whole Mafic Index (Figure 18e) which shows a steady increase in concentration of dark and heavy minerals towards the north-east. This result is consistent with the

conclusion of Bott et al. (1970), that the Bodmin Moor Granite has an increased density on its northern side. The ratio of potassium feldspar to plagioclase (Figure 18f) is highest in the same area as the tourmaline maximum, and occupies much of the granite in which Edmondson (1970) found microcline to be absent.

Figure 19a shows a well defined central area with specimens having Fe_2O_3 + MgO less than 1.8 per cent, there are two other specimens, one in the extreme north and one in the extreme west, which have low values of this parameter. Other chemical data for the coarse granites are plotted in Figure 20 and confirm that this central area has a composition distinct from its surroundings. Figure 19b shows that specimens from this area have low normative anorthite. Edmondson (1972) used trend-surface

★ **Sample preparation and geochemical analysis**

The original geochemical data presented was produced as follows: 1–2 kg specimens were trimmed to remove weathered material before being crushed to < 5 mm fragment size by hydraulic splitter followed by jaw-crushing. The material was then coned and quartered to give a fraction of 300–400 g which was powdered in a Tema tungsten carbide mill. Fractions of this powder, obtained by coning and quartering, were analysed by the methods then employed by the analysts named in the laboratories indicated in the tables.

The Keele University SiO_2 determinations were made by fusing the rock powder into discs with lithium metaborate and analysing by a Bausch and Lom/ARL 8420 X-ray fluorescence spectrometer using a rhodium tube, while loss on ignition was measured by the conventional method of weighing before and after heating to > 110°C. Samples sent to King's College, London, were analysed by inductively coupled plasma spectrometry, as described by Walsh (1982), and those sent to the Universities Research Reactor by neutron activation analysis (Duffield and Gilmore, 1979). X-ray fluorescence and atomic emission spectrometry were used in the Earth Resources Centre, Exeter.

Table 4 Chemical analyses of Bodmin Moor granites.

Number of samples	Ga 10	Gb 4	Gc 1	Gd 6	Ge 9	Gf 2
SiO$_2$	71.25 (70.17–72.85)	70.87 (70.54–71.12)	72.36	71.56 (70.20–72.53)	71.10 (70.28–72.38)	73.02 (72.83–73.21)
TiO$_2$	0.19 (0.11–0.28)	0.22 (0.12–0.28)	0.28	0.19 (0.11–0.25)	0.23 (0.18–0.29)	0.10 (0.07–0.12)
Al$_2$O$_3$	14.97 (13.95–16.81)	14.96 (14.00–15.58)	14.54	15.01 (14.31–15.62)	15.27 (14.75–15.82)	14.88 (14.45–15.31)
Fe$_2$O$_3$t	1.58 (1.07–2.18)	1.82 (1.20–2.16)	2.13	1.63 (1.09–1.99)	1.88 (1.46–2.18)	0.92 (0.88–0.96)
MnO	0.05 (0.03–0.06)	0.05 (0.04–0.06)	0.04	0.05 (0.04–0.06)	0.04 (0.03–0.05)	0.03
MgO	0.31 (0.17–0.67)	0.37 (0.20–0.46)	0.43	0.32 (0.17–0.44)	0.39 (0.27–0.47)	0.09 (0.02–0.16)
CaO	0.60 (0.40–0.88)	0.77 (0.57–0.96)	0.81	0.64 (0.44–0.72)	0.73 (0.46–0.98)	0.36 (0.25–0.46)
Na$_2$O	2.68 (2.20–3.02)	2.74 (2.07–3.14)	2.74	2.89 (2.49–3.17)	2.95 (2.47–3.37)	2.28 (2.10–2.46)
K$_2$O	5.37 (4.65–7.01)	5.63 (4.63–6.63)	4.73	5.37 (4.48–5.87)	5.30 (4.97–5.71)	6.18 (5.73–6.63)
P$_2$O$_5$	0.21 (0.19–0.23)	0.20 (0.19–0.21)	0.21	0.20 (0.17–0.23)	0.21 (0.19–0.23)	0.23 (0.18–0.27)
H$_2$O$^+$	1.05 (0.77–1.70)	0.87 (0.75–1.13)	0.95	0.87 (0.80–0.92)	0.98 (0.74–1.27)	1.21 (1.07–1.34)
Ba	154 (74–249)	161 (93–197)	145	138 (117–182)	195 (116–222)	141 (96–186)
Ce	49 (19–69)	56 (53–76)	80	49 (19–68)	61 (50–81)	11 (10–12)
Co	49 (28–68)	35 (27–48)	40	48 (34–61)	42 (31–60)	not det.
Cr	11 (6–18)	17 (13–19)	13	10 (7–13)	14 (9–18)	19 (8–30)
Cu	7 (4–19)	4 (3–6)	4	4 (3–4)	8 (3–38*)	6 (3–9)
La	25 (10–36)	30 (20–40)	40	25 (16–35)	32 (27–43)	9 (7–10)
Li	363 (111–608)	390 (209–510)	357	401 (314–597)	268 (85–512)	225 (145–305)
Nb	14 (11–17)	14 (12–14)	14	14 (13–16)	13 (12–15)	13
Ni	10 (7–14)	11 (8–16)	12	10 (8–12)	11 (8–15)	4 (2–5)
Pb	31 (21–50)	28 (24–30)	31	29 (27–33)	33 (27–37)	40 (29–51)
Rb	469 (343–591)	460 (441–502)	414	471 (439–545)	401 (336–465)	438 (416–460)
Sc	3 (3–4)	4 (3–4)	4	4 (3–4)	4 (3–4)	not det.
Sr	68 (32–101)	74 (54–97)	78	72 (62–90)	94 (71–107)	74 (41–106)
Th	13 (9–18)	16 (11–21)	24	12 (7–20)	15 (12–24)	3 (1 only)
U	9 (4–17)	9 (7–12)	5	9 (3–21)	9 (5–15)	8 (1 only)
V	12 (6–18)	14 (8–18)	19	12 (7–16)	14 (9–19)	6 (2–10)
Y	11 (9–13)	12 (11–13)	13	10 (9–12)	11 (10–13)	10 (9–10)
Zn	44 (23–85)	65 (36–99)	40	39 (30–47)	44 (36–77)	37 (33–41)
Zr	72 (48–96)	73 (55–85)	89	65 (45–88)	80 (72–88)	45 (34–55)
Normative:						
Q	34.14 (29.22–36.09)	31.70 (28.49–37.98)	36.20	32.92 (29.14–37.15)	31.83 (27.92–35.77)	36.08 (35.56–36.60)
Or	32.66 (29.03–41.95)	34.08 (28.88–40.03)	28.48	32.43 (28.07–34.74)	31.94 (30.22–34.22)	37.26 (34.53–39.99)
Ab	23.40 (19.26–26.68)	23.79 (17.89–26.88)	23.63	24.99 (21.77–26.94)	25.49 (21.53–28.65)	19.69 (18.14–21.23)
An	1.74 (0.72–2.96)	2.59 (1.64–3.50)	2.75	3.56 (1.16–2.33)	2.39 (1.07–3.60)	0.30 (0–0.60)

$^+$ Includes one specimen from near line of Cu vein.

t Total iron as Fe$_2$O$_3$.

Bracketed figures show component variation (per cent or parts per million) of samples within the granite type.

Unpublished data from C S Exley; Inductively coupled plasma analysis by J N Walsh, King's College, London (except SiO$_2$ and H$_2$O$^+$ respectively, X-ray fluorescence analysis by M Aikin and D W Emley, University of Keele; and Th and U, neutron activation analysis by G R Gilmore, Universities Research Reactor, Risley).

analysis to demonstrate the distribution of major and minor elements from specimens used in this study which also showed roughly concentric distributions.

Rare-earth-element compositions and distribution patterns from specimens of the eastern part of the Bodmin Moor Granite are consistent with those of the other Cornubian granites, and show the relatively high concentrations of light rare-earth element typical of high-level evolved granites (Table 5 and Figure 21).

The fine-grained granites tend to be richer in quartz, potassium feldspar and muscovite than the coarse-grained granites (Table 3), and are correspondingly rich in SiO_2, K_2O and P_2O_5 (Table 4). They tend to be poorer in biotite and its associated ferromagnesian elements. Like the coarse-grained granites, however, they show a wide range in composition (Figure 20), and although grouped with the 'inner' granites in Figure 20, individual analyses are widely scattered. They have rare-earth-element patterns similar to those of the coarse granites, but have a marked europium negative anomaly (Figure 21).

Petrogenesis

Brammall (1926), and Brammall and Harwood (1932), believed that the coarse-grained rocks of the Dartmoor granite pluton consisted of at least two intrusive phases, and this sequence was thought by Ghosh (1927) to be also present in the eastern part of the Bodmin Moor pluton. Although multiple intrusion is a feature of the Carnmenellis pluton (Ghosh, 1934; Al Turki and Stone, 1978; Leveridge, Holder and Goode, 1990), it is not now considered to have been a factor in the development of the Dartmoor or Bodmin Moor plutons. The variations noted by Brammall (1926) and Brammall and Harwood (1932) and by Ghosh (1927) are here interpreted as facies changes due to differentiation in situ, probably with some contamination by country rock. The composition of the south-west England batholith

varies along its length, the earlier coarse-grained part of the Bodmin Moor Granite differing from those of Dartmoor and St Austell granites. The concentrations of biotite and andalusite, around the margins of the Bodmin Moor Granite, suggest the incorporation of some pelitic sedimentary material, and the concentration of potassium feldspar close to the margins (except in the north-west) is consistent with this. It is also possible that, given the lower concentrations of potassium feldspar and normative anorthite in the central area, there was some migration of potassium to the outer, cooler parts of the pluton. The distribution of textural evidence provided by megacrysts in the present district is too irregular for firm conclusions to be drawn.

The origin of the megacrysts remains unresolved. The evidence suggests that they resulted from subsolidus metasomatism rather than magmatic growth. They enclose, commonly in a concentric zone, minerals of earlier magmatic origin, and have irregular margins which partially envelop adjacent crystals. Many have relatively well-developed crystal faces; others link with smaller, interstitial crystals or irregular areas of potassium feldspar, the whole being optically continuous. Intergrowths are common at the margins, and indicate reaction. These phenomena, and the variation in size, are not common where phenocrysts have grown in a magma, nor do orthoclase phenocrysts consistently occur in a matrix that contains either orthoclase or microcline.

Edmondson (1970) linked the irregular distribution of microcline in the Bodmin Moor Granite and exsolution in potassium feldspar to form microperthite with late-stage hydrothermal activity. He attributed the presence of orthoclase in preference to microcline to greater late-stage activity in the south of the pluton. This encouraged megacryst growth rather than the structural inversion of feldspar, so delaying the latter process until insufficient energy remained in the system for the inversion to be completed. Evidence of late-stage activity is provided by the concentration of tourmaline in the same area.

The chemical compositions of the fine-grained granites show them to be slightly evolved variants of the coarse-grained granite. The field evidence suggests that they intruded the coarse-grained granite, sending out sheets and veins that irregularly ramify through the adjacent rocks.

GRANITE AND APLITE VEINS

Veins of granite, aplogranite and aplite occur in the coarse- and fine-grained granites, and represent a range of late-magmatic activity. Although nearly all these rocks are technically leucogranites, with less than 5 per cent of dark minerals, they do not fall neatly into sharply defined categories. Some are aplogranites with little or no biotite.

Granite veins

Small granite veins, rarely exceeding a few tens of metres in length or 1 m in thickness, intrude both the granite

Table 5 Rare-earth-element concentrations (ppm).

Coarse-grained granite		Fine-grained granite	
La	36.4	La	12.6
Ce	86	Ce	29.2
Pr	not det.	Pr	not det.
Nd	34.2	Nd	12.7
Sm	5.8	Sm	2.8
Eu	1.7	Eu	0.43
Gd	6	Gd	3.3
Tb	not det.	Tb	not det.
Dy	3	Dy	2.2
Ho	not det.	Ho	not det.
Er	2.4	Er	0.65
Tm	not det.	Tm	not det.
Yd	1.13	Yb	0.55
Lu	0.17	Lu	0.07

Single sample determinations, unpublished data from C S Exley. Neutron activation analysis by G S Gilmore, Universities Research Reactor, Risley.

Figure 20
Variation
in some
constituents of
coarse-grained
granite (O) and
fine-grained gran-
ite (X).
a. Fe$_2$O$_3$ + MgO
v. CaO + K$_2$O +
Na$_2$O.
b. Ti × 100 v. Zr.
c. K/Rb v. Zr.
d. Y v. Zr.
e. Y v. Nb.
f. Ba v. Rb v. Sr.

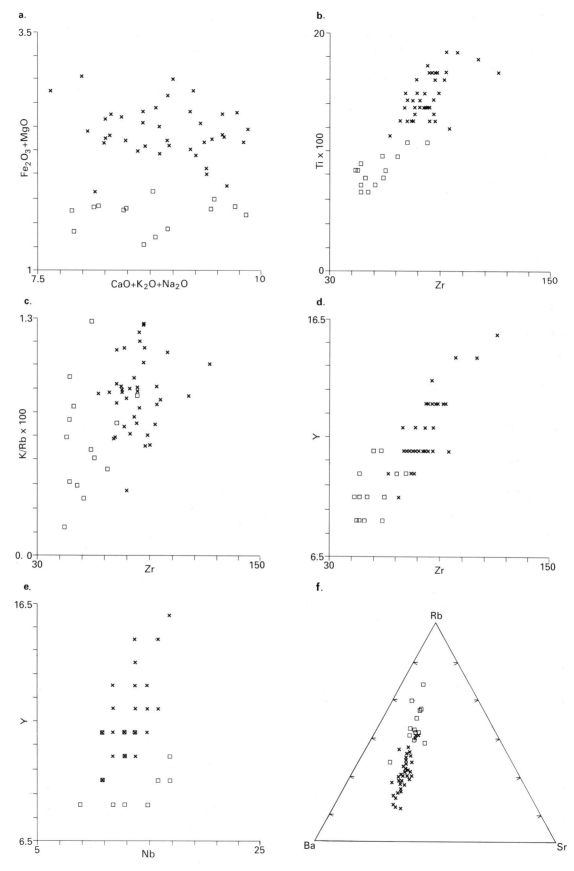

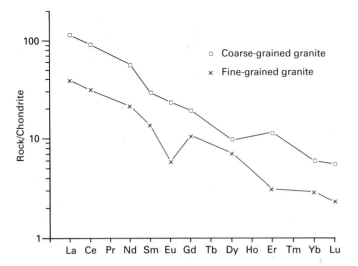

Figure 21 Chondrite-normalised rare-earth-element patterns for coarse-grained (O) and fine-grained (X) granites.

and the country rock and were described by Reid et al. (1910). Their composition is similar to the parent granite. Examples occur at Carbilly Tor [SX 1252 7545] and Row [SX 0940 7666] quarries. Between Rough Tor and Little Rough Tor [SX 1470 8085], similar rocks occur in a zone about 40 m wide. Although this may be an unusually wide vein the field relations are not seen and it may form part of a larger body of fine-grained granite.

Aplites and aplogranite veins

Aplite and related veins are not seen to arise from either of the main granite types, but cut both. They are commonly some hundreds of metres or a few kilometres in length but are generally only a few tens of centimetres wide. They vary greatly in attitude from vertical to near-horizontal. Their strike is not closely related to any of the principal joint directions, but some are offset by joints.

The aplites and aplogranites have an allotriomorphic, granular texture but not all are equigranular. Examples from Brown Willy [SX 1587 7998], Carbilly Tor [SX 1258 7549] and the new A30 road cutting [SX 1827 7682] immediately north of Jamaica Inn are poorly megacrystic; the inclusions within the megacrysts suggest that they contain crystals of two generations set in a more truly aplitic matrix. These megacrysts include: quartz, commonly in composite grains, angular to rounded with irregular margins and up to about 5 mm across; potassium feldspar both as anhedral grains, enclosing small quartz and plagioclase crystals, and as megacrysts up to 4 mm (these are the only megacrysts in the Carbilly Tor specimen); and plagioclase, subhedral to anhedral, normally zoned, up to 4 mm. Most megacrysts are 2 to 3 mm across. The matrix in these and in equigranular examples (such as those from Rough Tor [SX 1456 8070] and a boulder from Manor

Common [SX 1345 7430] contains crystals ranging in size between about 0.1 mm and 2 mm. These crystals are all anhedral, with the exception of a few plagioclase and mica which may be subhedral, and they are generally irregular in outline. They include quartz, potassium feldspar, plagioclase (more abundant than potassium feldspar in places), sporadic biotite in ragged flakes usually altered to chlorite, muscovite in excess of biotite, tourmaline and some irregular or rounded grains of apatite. The potassium feldspar in the Jamaica Inn, Rough Tor and Carbilly Tor specimens is microcline microperthite. The tourmaline is rounded, irregularly prismatic or reticulate and is commonly yellow-brown with irregular blue patches or displays sporadic zoning. Biotite and tourmaline tend to vary inversely throughout. Rare specimens of true aplite, with allotriomorphic equigranular textures, occur in association with pegmatites in a similar fashion to the occurrences at Megiliggar Rocks in the Penzance district [SW 610 265] (Goode and Taylor, 1988). Exposures of this type of rock in the Bodmin Moor granite are found in the De Lank [SX 1010 7550] (Plate 5b) and Hantergantick quarries. An exposure [SX 1030 7570] at the former locality shows a banded aplite vein 112 mm wide which passes into a pegmatite of approximately the same width. A 20 mm-wide zone between the two includes rock coarser than the aplite, and this appears to be disrupted and contorted and to include a 2 mm-wide zone of tourmaline crystals. The feldspars in the pegmatite are orientated with their long axes perpendicular to the aplite vein.

Petrogenesis

The microgranite veins have not been radiometrically dated, but they probably include examples from both episodes of magmatic activity. They may thus be direct descendants of both coarse- and fine-grained granite magmas. The extent to which they have penetrated their host rocks is a measure of the prevailing local physical and chemical conditions, and thus the mobility of the magma.

The aplites and aplogranites are more evolved rocks than the granites, and represent more advanced physico-chemical conditions associated with the concentration of volatile constituents at the end of each magmatic stage. These rocks are not usually associated with pegmatite, despite the occurrence noted above, and there is little evidence either of the separation of an aqueous fluid phase, such as that which gives rise to aplite-pegmatite complexes, or of the partition of potassium and sodium into different fluids. The ratio $K_2O:Na_2O$ is similar to that in the parent granite. The Bodmin Moor aplites may therefore be regarded as local expressions of the movement of mobile residual magma into zones of weakness and lower pressure.

Chemical analyses of two aplites from the district are given in Table 6. The specimen from Brown Willy [SX 1587 7998] has concentrations for several constituents outside the ranges used in Figure 20. However, those for $Fe_2O_3 + MgO$ and for Zr show that it, like the Bolventor speci-

Table 6 Chemical analyses of aplites from Bodmin Moor.

	a	b
SiO$_2$	74.0	74.79
TiO$_2$	0.10	0.09
Al$_2$O$_3$	14.85	14.21
Fe$_2$O$_3$	1.03	0.77
MnO	0.06	0.02
MgO	0.03	0.008
CaO	0.34	0.28
Na$_2$O	2.75	3.29
K$_2$O	5.09	5.21
P$_2$O$_5$	0.32	0.30
H$_2$O$^+$	1.26	0.62
Ba	44	33
Ce	5	2
Co	not det.	not det.
Cr	16	15
Cu	3	3
La	7	82
Li	472	17
Nb	19	2
Ni	4	57
Pb	19	449
Rb	664	not det.
Sc	0	24
Sr	21	11
Th	not det.	not det.
U	not det.	not det.
V	5	6
Y	9	46
Zn	54	34
Zr	43	not det.
Normative		
Q	38.69	35.96
Or	30.54	31.13
Ab	23.62	28.14
An	0	0

Total iron as Fe$_2$O$_3$; n.d. = not detected; not det. = not determined

Specimen a. from Bolventor [SX 1827 7682].
       b. from Brown Willy [SX 1587 7998].

Unpublished data from C S Exley. X-ray fluorescence analysis by M Aiken, University of Keele (except MgO and Li; the former X-ray fluorescence and the latter atomic emission spectrometry analysis by J Merefield, Earth Resources Centre, Exeter).

men [SX 1827 7682], is more highly evolved than both the 'inner' coarse-grained and fine-grained granites (Figure 20a, c).

HYDROTHERMAL ALTERATION OF THE GRANITE

Greisenisation

The alteration of feldspars to secondary mica and quartz, like tourmalinisation, occurred in two phases, of pre- and post-joint age. Greisenisation was caused by the separation of an aqueous fluid phase that contained fluorine and was sufficiently acid to attack the feldspar, but in which boron was not significant. The early phase is seen in most thin sections as the alteration, usually to a minor

degree, of potassium feldspar to secondary white mica. This rarely extends beyond strings of small flakes in the cleavages and around the edges of crystals; near-pseudomorphs in secondary mica and quartz are rare, and occur, for example, in specimens from Redhill Downs [SX 1641 7191] and near Crowdy Dam [SX 1399 8335]. The granite in both these localities is relatively soft.

Greisenisation of post-joint age occurs as veins in which the alteration to secondary mica and quartz is intense. Usually 10 to 20 mm wide, these veins often form the outer zones of quartz-tourmaline veins; they are widest in areas where kaolinisation has been most severe, for example at Stannon [SX 1280 8120] and Park [SX 1940 7080] china clay pits. The greisen veins trend approximately east–west.

Unusual developments of greisenisation which apparently predate the joint system occur at Brown Gelly [SX 1970 7250] and Showery Tor [SX 1490 8130]. There, resistant ridges that differ in trend from the joints by up to 30° stand 10 to 20 mm above the rock surface. In thin section, these are seen to contain minute veinlets of secondary mica linking greisened feldspars; it is thought that they were formed by fluids that penetrated zones of weakness after the magma had crystallised but before the rock was cold enough to have developed a joint system (Exley, 1961).

Tourmalinisation

The presence of tourmaline as an ubiquitous rock-forming mineral is evidence of the early role played by boron in the development of the Bodmin Moor granite. Brammall and Harwood (1925) distinguished between pre- and post-consolidation generations of the mineral, and separated the former into 'primary' and 'secondary' phases. They noted that very little of the tourmaline was truly primary or 'pyrogenic', the bulk being formed by 'autopneumatolysis'. They described a large number of modes of occurrence in the Dartmoor granites, ranging from minute primary grains, though irregular crystals replacing feldspar and biotite, to well-formed crystals in cavities and pseudomorphous aggregates. Similar varieties can be recognised in the other granites.

The pre- and early post-consolidation generations of tourmaline have been described as 'pre-jointing' (e.g. Exley and Stone, 1964), because they are related directly to the magma or fluids associated with it. Tourmaline of this age is yellowish brown in thin section with irregular zones and patches of blue. It is ubiquitous, but varies greatly in amount. Rounded nodules of tourmaline-rich granite in which the mineral has similar characteristics also occur. The tourmaline prisms in these nodules usually radiate from a centre to form 'suns'. Where replacement occurs, the textures vary from simple embayment to lacy developments of tourmaline that spread through pre-existing crystals.

The post-consolidation tourmaline occurs in joints, veins and cavities in the granite. Joint fillings and veins are usually only up to 10 mm in width; miarolitic cavities are mostly about 50 mm across, although some are much larger.

Tourmaline-filled veins, joints and faults strike roughly east–west; examples occur in the cutting below Colliford

Dam [SX 1794 7102], at Cabilla Tor [SX 1494 6976] and Colvannick Tor [SX 1255 7177]. Pegmatitic pods occur in a small tor [SX 1606 7323], 0.25 km north of Simon's Stone, at Carkees Tor [SX 1388 7633] and in the cutting [SX 1780 7585] on the east side of the A30 road 1 km south-west of Bolventor. In both veins and pods, the tourmaline is usually associated with quartz, and in the coarser, more pegmatitic pods feldspar and white mica are commonly also present. Such pods are up to 0.2 m across.

Fine-grained, acicular tourmaline commonly coats the movement surfaces of faults, and forms a cement between fragments in breccias. A good example of a tourmaline veneer occurs along the fault [SX 1010 7550] bounding the north side of De Lank Quarry and running through Hantergantick Quarry [SX 1034 7569]. Tourmaline in breccia is abundant in both the western [SX 1018 8010] and eastern [SX 1021 8024] crags of the Devil's Jump. It is black in hand specimen, but the post-consolidation or 'post-joint' tourmaline is blue-green in thin section, and is much more acicular in habit than the 'pre-joint' variety.

Quartz-tourmaline rock ('peach' or 'schorl rock') is not common on Bodmin Moor, although it is present in boulders on the lower ground, especially where there is evidence of china clay. The tourmaline in these rocks is also needle-like and tends to form a felted texture.

As temperatures fall in boron-rich magmas, an alkali-silica-boron-rich aqueous phase separates from the main silicate melt (Pichavant, 1979; and Charoy, 1982). Some of this, in the interstices of the already-crystalline phases of the magma, will crystallise as tourmaline or will replace existing minerals with tourmaline; some, in immiscible globules of varying size, will solidify in situ to form patches and pods of tourmaline or tourmaline-bearing pegmatite. If the granite is fractured, some of the boron-rich differentiate can escape through the fissures to crystallise as tourmaline veins. Power (1968) has shown not only that there are differences in the compositions of the magmatic and hydrothermal tourmalines, but also that variations in the former reflect the degree of fractionation of the magma. Manning (1991) has subsequently described the magmatic tourmaline as belonging to the schorl-dravite series, the hydrothermal tourmaline as being nearer to stoichiometric schorl, and the tourmaline from breccias in the metamorphic aureoles as approaching uvite in composition. The Bodmin Moor Granite contains examples of all three of these, though not so spectacularly as the St Austell and Land's End granites.

Kaolinisation

Kaolinisation is the conversion of plagioclase by acid, aqueous fluids to an aggregate of clay and secondary white mica. The process starts in the cores of the crystals, which are more anorthitic, and progresses outwards. In its extreme stages, all the original minerals of the rock are altered to some degree: quartz is partly recrystallised, biotite loses iron and becomes bleached, and tourmaline is corroded. Two types of kaolinisation have been recorded: surface, shallow alteration due to percolating rainwater, possibly initiated in the tropical conditions of the Tertiary, and deep-seated alteration.

Alteration of the feldspars to clay minerals gives rise to a soft, friable rock which is easily eroded. In the district, this alteration is largely responsible for the undulating surface of Bodmin Moor where resistant granite forms the tors and hills, and the valleys and depressions are underlain by a mixture of detritus washed from the granite and kaolinised granite. This type of widespread kaolinisation extends only a few metres down into the granite. Cross-sections can be seen in many road cuttings and around the perimeters of china clay pits.

The second type of kaolinisation is related to deep-seated fracture systems, and follows those which traverse the pluton and delineate much of the drainage pattern of the district. The most intense alteration occurs where fracture systems intersect, as at Stannon and Park pits, but is not confined to such areas. Former clay pits [SX 1360 7322: SX 1365 7307: SX 1440 7178] in and near the upper parts of the Warleggan River and at Hawkstor [SX 1500 7450] show the same feature. The Hawkstor pit is on the line of a major east–west fault zone which was clearly visible when the pit was worked.

Of the various types of alteration described above, greisenisation and tourmalinisation can be linked directly to the immediate post-emplacement cooling history of the host pluton (Shepherd et al., 1985, Chesley et al., 1993). Though no detailed studies have been carried out on the age of the Bodmin Moor Granite mineralisation, the relationships described in Chapter Eight (p.85) suggest that, in the case of tin-bearing veins within the granite, the deposition of vein tourmaline is later than the formation of greisen.

GRANITE JOINTS

The jointing of the Bodmin Moor Granite comprises a subhorizontal set, the spacing of which increases with depth and two roughly vertical sets trending north-west and east-north-east intersect at angles between 70° and 90°. Individual subvertical joints may be replaced by conjugate pairs, and there are zones where some movement has occurred. Exley (1965) noted that movements along vertical joints separates the Bodmin Moor Granite into large blocks; some of these movements, notably those in north-west-trending zones are related to major structural features such as the Portwrinkle and Portnadler faults. Two of the east-north-east-trending 'belts' of Exley separate the northern and southern flanks of the granite from an uplifted central area. The cause of the uplift is unknown, but Bristow (1990) has suggested that it might relate to an intrusion of Li-mica granite at depth.

CONTACT METAMORPHISM

The outer limit of the metamorphic aureole enclosing the Bodmin Moor Granite is taken in the present work at the appearance of spotting in pelitic lithologies. There are considerable variations in the geochemistry of the country rocks of the aureole. Two main rock types have been recognised, grey porphyroblastic, mica schists and, less commonly, dark flinty hornfelses known locally as calcflintas.

The width of the metamorphic aureole reflects the dip of the granite margin and the lithologies of the country rocks. The aureole extends for 0.8 to 1.5 km in the non-calcareous pelites, but may exceed 2.5 km in the calcareous pelites. The outcrop of the aureole is narrow along the north-western margin of the granite; a borehole [SX 1635 8277] at Oldpark shows the granite contact to be faulted and to dip 70° to 75° north. A spur of the granite at Lower Moor, north of the Stannon fault, has an aureole width of 2 to 3 km, and suggests that the contact between the granite and the country rock dips at a shallow angle beneath Camelford. The fault there has probably preserved an area of the granite that formed close to the roof of the pluton.

The western margin of the granite is mostly vertical and in places cut by faults of the Cardinham fault zone. The uniform width (1 km) of the aureole suggests that no significant thickness of county rock has been cut out by faulting. The intrusion appears to have exploited pre-existing fractures in the fault zones; some of these were reactivated during and after intrusion.

The greenish grey slates of the Tredorn Slate Formation are altered in the aureole to silver-grey mica schists. These are characterised by an abundance of white mica and show a greenish grey spotting, usually elongated parallel to the foliation. The spots, which commonly reach more than 5 mm in diameter, are formed of finely crystalline aggregates of white mica and chlorite. Porphyroblasts of chloritoid and andalusite commonly cut across the foliation of the rock, and may be associated with small patches of iron oxide and prisms of rutile. Clusters of andalusite needles, up to 30 mm long, locally occur in darker grey schists at Camelford [SX 1051 8365] and Tyland Road [SX 1202 8404]. Cordierite is not characteristic of this lithology.

Dark grey mica-schists are formed from Trevose Slate. They carry a higher proportion of chlorite and finely divided iron oxide, but less mica than the silver-grey schists. Muscovite and biotite are present in approximately equal quantities; close to the granite the slates assume a brown tinge and become more coarsely crystalline as biotite replaces chlorite. Contact metamorphism is characterised by cordierite spotting. A progressive increase in metamorphism occurs as the granite is approached, from weakly recrystallised cordierite slates with white mica enhancing the foliation, through cordierite-andalusite slate and phyllite with weak fabric-parallel segregations of quartz, to cordierite-andalusite-biotite phyllite and schist. In the last-named lithology, extensive secondary white mica is developed and there is much evidence of tourmalinisation, commonly as a result of emanations from vein systems that caused the replacement of earlier contact minerals.

Dark flinty hornfelses, the product of contact metamorphism of impure calcareous rocks, are interbedded with the dark grey micaceous schists. The hornfelses are brittle and dense and commonly show compositional banding of granular and compact calcflintas. Reid et al. (1910) described such lithologies in the Tredorn Slate at a ford east of Camelford [SX 120 840], at Trethin [SX 105 819], and in sections in the River Camel between Helsbury [SX 092 784] and Trecarne [SX 096 806]. The granular calcflintas is characterised by a coarse ground-mass of pink garnet or yellowish brown idocrase carrying many small grains of pale green pyroxene; the compact calcflintas contain mainly green pyroxene with some garnet and idocrase and subordinate hornblende, white mica, sphene, epidote, clinozoisite and calcite. Comparable calcflintas occur in the Trevose Slate, for example at Penrose [SX 0865 7629], and in the Rosenun Slate at Pantersbridge [SX 1592 6804].

Metabasite mineral assemblages adjacent to the granite margin indicate hornblende-hornfels facies metamorphism. The mineralogy comprises quartz, albite, hornblende, prehnite, biotite, calcite and sphene with the opaque minerals pyrite, pyrrhotite, chalcopyrite and arsenopyrite. Hornblende is developed within planes corresponding to S_1 in the country rocks; it may show a preferred orientation and impart a lineation to the rock fabric. Disseminated sulphide minerals, principally pyrrhotite, are characteristic of all the aureole metabasites.

Calcflintas containing much alkali feldspar and modified adinole structure enclose most of the metabasites in the district. Reid et al. (1910) recorded instances where the adinoles were tectonically disturbed prior to contact metamorphism; this suggests a complex history of formation.

FIVE

Minor igneous intrusions

Two suites of dykes, probably of different ages, occur in the district. Acid porphyrites, commonly known as elvans, consist of quartz-feldspar porphyry, they intrude both the county rocks and the granite. Lamprophyres occur only in the country rock outside the thermal aureole of the granite.

ACID PORPHYRITE DYKES (ELVANS)

Quartz porphyry dykes are widespread in Cornubia, and are clearly associated with the granite batholith. The dykes vary in width from about 0.5 m to about 40 m and dip at angles steeper than 60°, although more sill-like bodies occur, for example near Lanteglos [SX 0894 8210]. The elvan dykes in this district are mostly east–west-trending, but there are some variations. Dykes east and south of Camelford [SX 1143 8325] and south of Lanteglos and St Teath [SX 0593 7889] strike north-eastwards. These separate outcrops may be part of one long dyke, which continues from west of St Kew [SX 0040 7627]

There are more elvans in the south of the district than in the north. A group of dykes occurs in the Helland–St Mabyn–Blisland area, and a swarm of dykes occurs near Warleggan and Cardinham. Stream exposures [SX 1646 6898 and SX 1632 6923] east of Warleggan and in Treveddoe Wood [SX 1485 6924] were formerly mapped as a single, bifurcating 'sill-like elvan' (Reid et al., 1910, p.60); they are here regarded as separate intrusions. Good exposures of contact relationships occur in the many quarries in the elvans. At De Lank Quarry [SX 1000 7535] and two former quarries [SX 1010 7527; SX 1010 7520] to the south, the dykes are near-vertical. A quarry at Temple Tor [SX 1395 7338] exposes a 10 m-wide dyke dipping at 80° to the north. A westerly continuation of this dyke was formerly exposed close to the old road [SX 1339 7322]. The quarries at De Lank and Temple Tor retain a skin of felsitic-textured rock from the chilled margin of the dyke, and demonstrate an irregular, stepped surface where the magma had stoped its way into the host granite. At Temple Tor, much of the south-eastern face has a tourmaline vein left as a veneer. Such veins are common along faults in the Bodmin Moor granite, and this occurrence may indicate either that a fault provided the initial plane of weakness along which the elvan was intruded or that the movement took place after emplacement of the dyke. An exposure [SX 1290 7110] between St Bellarmin's and Colvannick tors shows elvan and granite interleaved.

At Blakes Keiro [SX 9619 7621], 1 km south-south-west of St Minver an elvan has an irregular, cross-cutting contact with the Polzeath Slate, fragments of which occur as inclusions in the dyke. The slates are thermally metamorphosed for a short distance on either side of the elvan.

Petrography

Acid porphyrites are light grey, some show pinkish or greenish tints. The concentration of phenocrysts is variable, ranging from a few per cent up to about 25 per cent by volume. The most abundant are potassium feldspar, mostly as subhedral crystals with irregular margins, but they may also be euhedral or anhedral. Their sizes are variable. The largest are 10 to 20 mm long, but most are about 6 mm; some are composite grains. Many are microperthitic, and many enclose small crystals of quartz, plagioclase or mica which may be zonally arranged. Micrographic intergrowth with quartz is not uncommon around the margins. Second in abundance is quartz, the majority of which forms rounded crystals or aggregates, many of which are embayed or corroded. These phenocrysts are smaller than those of potassium feldspar, usually 1 to 2 mm across. Quartz phenocrysts in elvans are often described in the literature as being bipyramidal, but such crystals are rare in the district. Some of the quartz phenocrysts have micrographic reaction rims. Plagioclase is commonly, present as phenocrysts up to 3 to 4 mm long which are subhedral with irregular margins. They are usually normally zoned and are commonly altered to secondary mica. Phenocrysts of muscovite, tourmaline and biotite (mostly chloritised) also occur. The Hobb's Hill elvan [SX 1851 6942], contains 3 mm-long pseudomorphs in chloritic minerals which have replaced other, now unidentifiable, mafic minerals. Polymineralic aggregates of quartz and feldspar occur as rounded, angular or irregular grains of similar size to the phenocrysts. They are almost certainly xenocrystic or xenolithic. Patches of pinite replacing cordierite are present, but rare.

The groundmass grain size of the elvans varies from almost cryptocrystalline to about 0.5 mm, but is usually about 0.1 to 0.2 mm. The groundmass consists of finely intergrown quartz, potassium feldspar and plagioclase which, in the coarser rocks, is roughly equigranular and allotrimorphic. Quartz and potassium feldspar are the most abundant minerals and textures vary from granophyric to micrographic. A more granular, less intergrown, matrix occurs in an elvan [SX 1293 8391] east of Camelford and at Pound's Cawnse [SX 1182 7134]. Plagioclase, biotite and tourmaline are rarely present; muscovite is ubiquitous. The presence of potassium feldspar and muscovite is reflected in the chemistry of elvans, which are rich in potassium.

Some of the elvan dykes in the district show laminar structure. Elsewhere in west Cornwall laminar structures indicate that the magma underwent shear as a result of viscous flow, or that it was intruded in more than one phase, or possibly a combination of both processes (Stone, 1968; Goode, 1973; Leveridge et al., 1990). Within the district, the elvans display considerable variation in grain size, from fine-grained chilled margins with few phenocrysts to coarser, commonly strongly porphyritic textures, in the central parts of dykes. The width of the coarser facies depends on the width of the dyke. The Temple Tor dyke has chilled margins about 50 mm wide with a few phenocrysts up to 2 to 3 mm long, slightly coarser zones 1.5 to 2 m wide in which the phenocrysts, though still few, are 5 to 20 mm long, and a central zone in which there are many phenocrysts in a groundmass of 0.5 to 1 mm grain size.

The elvans have been subjected to the same alteration processes as the granite. Tourmaline occurs as a replacement mineral, chiefly after feldspar and mica; greisenisation has caused the feldspar to break down to secondary mica, and kaolinisation has produced clay minerals and secondary mica after feldspar. Of the three processes, the last is the most common and reaches high levels of intensity where hydrothermal alteration has occurred.

Age

Throughout south-west England elvan dykes intrude coarse-grained granites, but not fine-grained granites. There is no field evidence as to the age of the elvan dykes in the district. In places they cut through mineralised veins. Radiometric dating by Rb/Sr methods of three elvan dykes at Brannel [SW 9553 5180], South Crofty mine, Pool, and Wherry, Penzance [SW 470 293] (Hawkes et al, 1975) gave a mean age of 269 ± 8 Ma; Darbyshire and Shepherd (1985) re-analysed the first two of these using the same method and obtained ages of 282 ± 6 and 270 ± 9 Ma respectively.

Petrogenesis

The radiometric dates suggest that the elvans formed during the second stage of granite magmatism. However, their textural and compositional heterogeneity suggests that their origin is complex. Stone and Exley (1978) have shown that some of the elvans are related to the coarse-grained, biotite granite and others to the finer-grained granites; Hawkes et al. (1975) have suggested that $^{87}Sr/^{86}Sr$ initial ratios of the elvans indicate a heterogeneous source. The xenolithic elements and reaction relationships described above are indicative of at least two generations of crystallisation. Together, these characteristics have led to the concept of the upward transport of solid granite fragments and early crystals in a gaseous fluid, giving a 'fluidised' system with reaction taking place between the two phases (Stone, 1968; Goode, 1973; Henley, 1974; Hawkes et al., 1975). Stone (1968) considered that solid fragments rich in potassium feldspar, and ion exchange during emplacement, could account for the potassium enrichment of elvans relative to their associated granites. Henley (1974), however, argued for reaction between solid granite and a residual, aqueous, potassium-rich fluid. There is general agreement that the flow structures and the variations in grain size in the elvan dykes have resulted from

changes in pressure and hence in the viscosity of the magma.

Table 7 gives a chemical analysis of the elvan at Temple Tor [SX 1395 7338] which, when compared with Figure 20, suggests that the elvan may be related to the coarse-grained granite.

LAMPROPHYRE (MINETTE)

Lamprophyric dykes are common in Cornwall and west Devon. They are closely related spatially and temporally to the granites. They are less common than elvans, and there are few in the district. They are generally 1 to 1.5 m thick, but range up to 7 m.

Table 7 Chemical analyses of Temple Tor elvan.

SiO_2	71.51
TiO_2	0.09
Al_2O_3	15.14
Fe_2O_3	0.77
MnO	0.05
MgO	0.09
CaO	0.69
Na_2O	2.20
K_2O	6.25
P_2O_5	0.27
H_2O^+	1.07
Ba	376
Ce	78
Co	n.d.
Cr	19
Cu	6
La	24
Li	231
Nb	13
Ni	5
Pb	32
Rb	409
Sc	n.d.
Sr	122
Th	not det.
U	not det.
V	35
Y	12
Zn	38
Zr	149
Normative	
Q	33.40
Or	37.48
Ab	18.18
An	0

Total iron as Fe_2O_3; n.d. = not detected; not det. = not determined

Unpublished data from C S Exley. X-ray fluorescence analysis by M Aiken, University of Keele (except MgO and Li; the former X-ray fluorescence and the latter atomic emission spectrometry analysis by J Merefield, Earth Resoures Centre, Exeter). See p. 46 for sample preparation and analytical methods.

Two exposures of lamprophyre occur on the north coast near Tregardock Beach; both show an east–west trend. The northerly dyke, at Minehousedoor Cove [SX 0415 8398], is 1 m thick; the southerly exposure, at Trerubies Cove [SX 0394 8397], shows several thinner, vertical dykes striking in the same direction. All appear to be following pre-existing joints or fault directions, and to cut across the cleavage of the adjacent rocks. An exposure [SX 945 743] at Cant Hill on the north bank of the River Camel, consists of two outcrops of variable inclination and thickness in which the rocks are vesicular and very altered. A former exposure on the Camel Estuary at Halwyn (Reid et al., 1910), is no longer visible.

An impressive exposure of lamprophyre occurs in an abandoned railway cutting [SX 0230 7310] near Lemail where two dykes, about 2 m and 7 m thick respectively, strike east–west and dip at about 45° N. They have irregular contacts and cut across the cleavage in the sediments which they have thermally metamorphosed for a distance of about 1.5 m. The rocks are fresh, relatively hard, dark purplish grey with abundant biotite both as phenocrysts and in the matrix, together with some porphyritic feldspar. The texture varies from fine-grained equigranular at the margins to slightly coarser and porphyritic in the middle part of the dykes. The phenocrysts include orthoclase, some as composite crystals, and biotite. In the groundmass, orthoclase is subhedral to anhedral and the biotite euhedral to subhedral, up to 0.5 mm long, roughly aligned and commonly with the dark margins typical of phlogopitic compositions. Small amounts of other mafic minerals, now replaced by chloritic pseudomorphs, are present as phenocrysts. Their shapes suggest that these were probably originally olivine and pyroxene crystals. Apatite occurs as anhedral, cloudy crystals and smaller euhedral needles; quartz and calcite are present interstitially. There is some opaque iron oxide, usually as an alteration product. Chemically, this rock is rich in TiO_2, MgO, K_2O, P_2O_5, Nb, Zr, Th and light rare-earth- elements (Table 8).

Age

The Cornubian lamprophyre intrusions are generally regarded as postdating the granite, although there is no field evidence to confirm this. The only published radiometric age (K-Ar) is 295 ± 2.6 Ma (Hawkes, 1981) which is similar to that of the oldest granite.

Petrogenesis

There is much uncertainty concerning the origin of the lamprophyre magma. Its chemistry suggests that it was derived from both mantle and crustal sources. Early views, that contamination of basaltic magma by pelitic crustal material or granite would explain its composition, have been thrown into doubt by analyses of the trace elements in the lamprophyre dykes (in particular Nb, Zr, Rb and Rh) and the ratios of these elements in the possible source rocks. Partial melting of the mantle and subsequent modification by crustal material has been shown to be inadequate to explain the observed trace-element ratios. Exley et al. (1983) have suggested partial melting of heterogeneous mantle material containing phlogopite, followed by metasomatic enrichment in incompatible elements, to explain the origin. Leat et al. (1987) have suggested that a large volume of gabbroic or lamprophyric magma might be present at depth and that the granites, basalts and lamprophyres could have been produced by varying degrees of fractionation. No geophysical evidence has yet indicated the existence of such a high-density body beneath the batholith.

Table 8 Chemical analyses of the Lemail lamprophyre.

	Weight %
N	5
SiO_2	51.18
TiO_2	1.95
Al_2O_3	13.68
$Fe_2O_3{}^t$	6.30
MnO	0.103
MgO	7.52
CaO	4.57
Na_2O	2.13
K_2O	7.79
P_2O_5	1.90
Total*	97.12
LOI	5.88
	ppm
Ba	5086
Be	10.1
Cr	238
Hf	35
Nb	54
Ni	255
Rb	262
Sr	1064
Ta	2.6
Th	51
U	10.0
W	4.2
Y	30
Zr	1420
La	261
Ce	479
Nd	224
Sm	25
Eu	5.63
Gd	—
Tb	—
Ho	1.39
Yb	1.23
Lu	0.12

t Total iron as Fe_2O_3

* Major elements determined on volatile-free basis; additional loss of volatiles (F, Cl, S?) during fusion.

N = number of analyses of different glass discs.
From Leat et al., 1987.

SIX

Structure

Four structural units (Figure 22) are juxtaposed in the district: the Tintagel, Bounds Cliff, Padstow and Warleggan units. These units are named after the successions which comprise them, with the exception of the Warleggan Unit which relates to the Liskeard Succession that crops out extensively in the adjacent Plymouth district. The geological structure of the district has been much discussed both in the past (see Chapter One for details) and more recently with the publication of a stratigraphical revision by Austin et al. (1992) and Selwood et al. (1993).

TINTAGEL UNIT

The Tintagel Unit occurs at the highest structural level in the region; overriding the Bounds Cliff and Padstow units on a complex of gently north-dipping structures showing both thrust and extensional features. It constitutes the Tredorn Nappe of Selwood and Thomas (1986a), and includes the southern development of the Tintagel High Strain Zone (Sanderson, 1979) of north Cornwall. The rocks, which form part of the Tintagel Succession, characteristically show epizone metamorphism (p.79).

FIRST-PHASE DEFORMATION (D_1)

D_1 is a ductile folding event generated by north–south compression which gave rise to the dominant gently dipping slaty cleavage (S_1) observed throughout the district. This strongly developed cleavage is axial planar to mesoscopic F_1 folds that are usually tight to isoclinal and initially faced southwards. Using way-up criteria, cleavage-bedding geometry and fold vergence, the position of

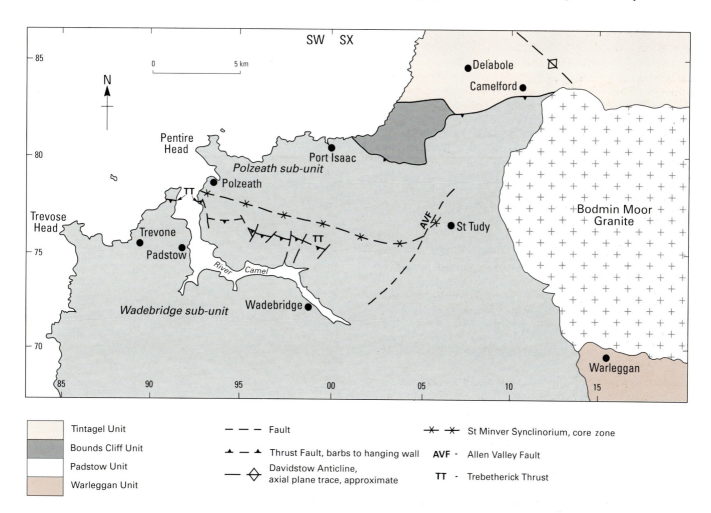

Figure 22 Distribution of structural units in the district.

sections on major fold structures has been identified (Warr, 1988). The gently north-dipping cleavage changes from being anticlockwise of bedding on the right-way-up limbs to clockwise of bedding on the overturned limbs, when viewed from the west. Using this criterion, approximately 70 per cent of the rocks within the coastal section appear to be inverted, representing the long overturned limbs of large-scale, asymmetric, south-facing folds.

Although D_1 strain was largely rotational, and the episode was predominantly a folding event, some localised thrusting did occur. All of these early structures were modified and overprinted by D_2 folding and thrusting events.

SECOND-PHASE DEFORMATION (D_2)

D_2 consists of a thrusting and shear-folding episode directed towards the north-north-west, causing thrust repetition and the dismembering of F_1 structures. The nature and distribution of D_2 structures across the area varies; shear folds are recorded and a D_2 crenulation cleavage is developed locally. All D_2 structures result from intense shear strain associated with north-north-west transport.

Two stages of D_2 deformation have been recognised, with a transition between them (Andrews et al., 1988; Warr, 1988); an early ductile deformation which was synchronous with greenschist-facies metamorphism, and a late brittle thrusting.

During early D_2, heterogeneous shear strain towards the north-north-west led to the development of the Davidstow Anticline (Warr, 1989) plunging 10° towards 330°, parallel to the trend of the regional stretching lineation, and was accompanied by large-scale penetrative deformation throughout the unit. Because the stretching lineation is consistent in orientation on both limbs of the Davidstow Anticline, it is suggested to have developed synchronously with it. Poorly developed alignments of andalusite and cordierite porphyroblasts in the aureole of the Bodmin Moor Granite, for example south of Slaughterbridge [SX 1115 8466], have the same trend. Although this might suggest that the contact metamorphism of the Bodmin Moor Granite was partly synchronous with early D_2 deformation, the orientation of these contact metamorphic minerals equally may have been controlled by pre-existing tectonic mineral lineations. The dip of S_1 and the orientation of F_1 folds in this area, reflect their position in the Davidstow Anticline. The gently north-dipping S_1 reorientated around the nose of the Anticline to dip south-westwards throughout the coastal section. D_1 folds in this area are thus gently reclined, with fold axes plunging west-south-west; sections oblique to these reorientated folds may show closures indicating either gently upward or downward fold-facing in a southerly direction on S_1. Such observations are a reflection of the plunge of reorientated folds, and not therefore of direct significance to the interpretation of D_1 deformation.

In the closure of the Davidstow Anticline, F_1 fold axes have been rotated from their original east–west trend to a north-north-west trend, parallel to the stretching lineation. The high strain involved in this early D_2

stretching also contributed to a high degree of sheathing of F_1 folds; the noses of most of the folds are sheared out, with fold closures towards the north-east or south-west being most common.

Since the D_2 strain was irrotational and broadly coaxial to D_1, most F_1 structures were tightened and extended rather than refolded during D_2, and a new element of stretching occurred in the north-north-west direction which initiated shearing and boudinage, well exposed at Trebarwith Strand [SX 048 864]. Although most strain in these rocks occurred during D_2, the observed sheathing of folds is the result of both D_1 and early D_2 strain components. The amount of curvature of F_1 fold axes before the later modification is unknown. The total amount of strain which occurred during D_1/early D_2 shear within the Tintagel Volcanic Formation at Trebarwith Strand shows typical X:Y:Z ratios for volcanic clasts of 2:1:0.2 (Andrews et al., 1988), indicating a high flattening strain.

Because much of the D_2 ductile deformation is approximately coaxial with D_1, S_1 and S_2 are effectively coplanar. Shear towards the north-north-west in D_2 thus took place along S_1 planes producing a composite S_1/early S_2 slaty fabric. This shear also produced local crenulation and pressure solution effects in the cores of tight F_1 and F_2 folds in the more arenaceous Lower Carboniferous lithologies at Tregardock Beach (Figure 23). Within argillaceous lithologies on the limbs of F_2 folds, however, S_2 acquires a slaty appearance. Locally, early ductile D_2 shearing along S_1 was intense enough to form tight, north-west-verging shear folds, with well-developed linear extension fabrics. Refolding of D_1 structures is also reported from Tregardock Beach [SX 0410 8417] (Andrews et al., 1988) and from the Tintagel Volcanic Formation near St Clear, for example at [SX 1963 8477] in which S_1 is transposed by S_2. Tight northward-verging F_2 folds within the Barras Nose Formation at Tregardock Beach [SX 0412 8412, SX 0416 8431] are associated with horizontal to gently south-dipping, northward-transporting thrusts (Figure 23).

Late D_2 brittle structures are only sparsely developed. The largest defines the base of the Tintagel Unit north

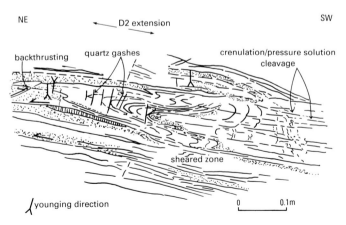

Figure 23 Section through a tight F_2 fold with pressure solution and crenulation effects in the core; Barras Nose Formation, Tregardock Beach [SX 0412 8412].

of Jacket's Point [SX 034 831] where it carries Tredorn Slate over Jacket's Point Formation. Planes of brittle shear, associated with the displacement in the hanging wall of this thrust, show overriding shear towards the north-north-west (Figure 24). Kink bands appear in the Riedel conjugate position.

THIRD-PHASE DEFORMATION (D_3)

D_3 consists of a phase of extensional faulting developed north of Minehousedoor Cove [SX 0416 8398] and is well displayed in the coastal section at Tregardock Beach [SX 0415 8438]. The faults, which may show gouges up to 1 m thick, dip gently (less than 30°) north-west, and commonly separate Upper Devonian rocks in the hanging wall from Lower Carboniferous rocks in the footwall (Figure 25). This relationship has arisen by low-angle faulting of strata that were already overturned by D_1 folding. Fault striations, shear fractures, vein arrays and kink bands are indicative of hanging wall movement towards the north-north-west; and fault planes cut down through the sequence, across the D_2 structures. North-verging folds are locally developed in the footwalls of the faults at Tregardock Beach and these are commonly transected by low-angle faults which may indicate sticking of basal fault surfaces during extensional movements.

The relationship between the D_3 extensional faults and the Davidstow Anticline is anomalous. These faults cut late D_2 thrusts, and yet appear to swing round the axis of the Davidstow Anticline with the regional strike (Warr, 1991; Plate 6a). They possibly represent a late phase of brittle gravity collapse. The chronology established by Warr (1989) highlights the timing and significance of the late D_2 thrusting (see p. 59) in what

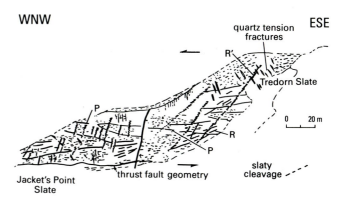

Figure 24 D_2 brittle shear fracture geometries at Jacket's Point [SX 071 834]. R = Riedel shear, R' = conjugate Riedel shear, P = thrust shear.

otherwise appears to be a continuum of first ductile compressive (D_1 and D_2), and then brittle extensional (D_3) events.

BOUNDS CLIFF UNIT

The Bounds Cliff Unit (Figure 22), comprising rocks of the Bounds Cliff Succession, is the lowest structural level recognised in the district; it was overridden during D_2 by the Tintagel Unit to the north-east and the Padstow Unit to the south.

FIRST-PHASE DEFORMATION (D_1)

D_1 was predominantly a folding event, producing large-scale tight folds with a well-developed axial-planar slaty

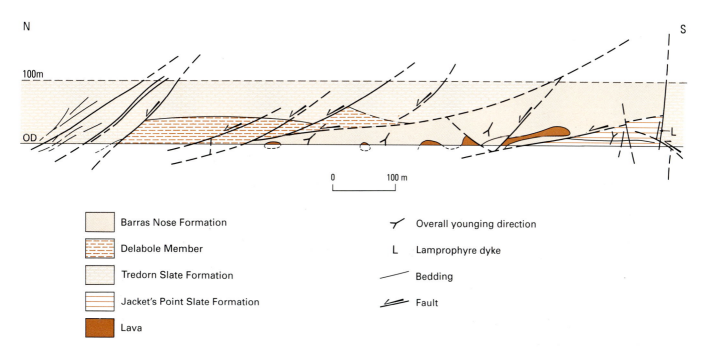

Figure 25 Structural section of Tregardock Beach, showing a series of northward-dipping D_3 extensional faults.

Plate 6a Trebetherick Thrust [SW 9265 7770] viewed from the west, in Daymer Bay (north).

This fault defines the Padstow Confrontation; it separates south-facing structures in the Polzeath Slate Formation of the footwall (to the left) from north-facing structures in the Harbour Cove Formation of the hanging wall (to the right). A thin development of windblown sand caps raised beach platform cutting angular head at the top of the cliff. (A15396)

Plate 6b Cleavage interaction at Trebetherick Point [SW 9251 7796] with interbedded purple and green slate of Polzeath Slate Formation showing well-developed cleavage relationships.

Two cleavages are present; steep (S_{1n}) cleavage dips to the north in the purple beds and to the south in the green beds, and a flat lying (S_{2n}) cleavage. The chevron folds of S_{1n} are related to S_{2n} and are hinged on bedding planes. (A15407)

Plate 6c Extensional fault and related folding at Port William [SX 0470 8629], Trebarwith Strand.

Drag folds, in the hanging wall of a low-angle normal fault in the Trambley Cove Formation, are subhorizontal and verge to the south-west. A crenulation cleavage is associated with the folds. The fault plane shows minor brecciation and an infilling of quartz. The rocks in the background are grey-green slates of the Tredorn Slate Formation. (A15390)

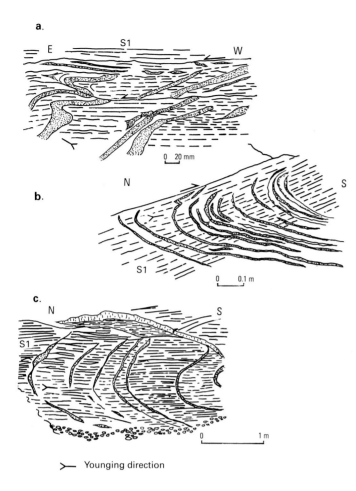

Figure 26 Small-scale, south-facing, F_1 folds in the Jacket's Point Slate. a. and b. Jacket's Point [SX 023 815], c. Barrett's Zawn [SX 027 819].

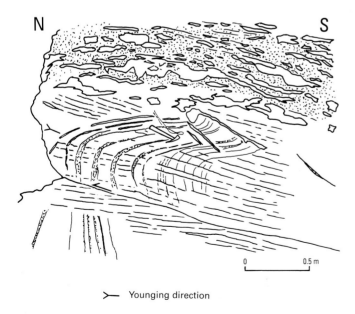

Figure 27 Profile section of a south-facing F_1 fold, transected by an early D_2 thrust in Jacket's Point-Slate, Bounds Cliff [SX 020 813].

cleavage (S_1). The 2.5 km of inverted Jacket's Point Slate from north of Jacket's Point [SX 0368 8346] to Bounds Cliff [SX 023 814], form the long overturned limb of a large-scale, gently inclined, asymmetrical, south-facing fold closing to the south. Within the hinge zone, at the southern end of the section [SX 0222 8144 to SX 0197 8125], symmetrical meso-scale folds with variable interlimb angles have fold axes plunging to the west-south-west. The S_1 cleavage dips between 2° and 32° to the south or south-east with downwards, southerly facing on S_1. At Pigeon's Cove [SX 017 812], F_1 folds face downwards towards the east-south-east with axes plunging 20° towards N017°. Northwards from Crook-moyle Rock [SX 0322 8277], there is a significant change in the facing of the unit from downwards and southwards with F_1 fold axes plunging 8° towards N132° to upwards towards the south-west (Figure 26).

SECOND-PHASE DEFORMATION (D_2)

D_2 structures, resulting from ductile shear and thrusting directed towards the north-north-west, are variably developed within the unit. Deformation, which was broadly coplanar with D_1, first took the form of a pervasive ductile shear, largely along S_1 planes.

Although most D_1 structures are not reorientated, ductile shearing along S_1 produced localised north-verging shear folds and associated shear indicators. Slick-encryst lineations, the product of the growth of extensional fibres of quartz, chlorite and iron sulphide minerals on S_1 planes and fault surfaces, are particularly common. The asymmetry of these fibres indicates hanging wall transport in a north-north-westerly direction.

Gently south-dipping, small-scale thrusts, showing little sign of fault-rock or gouge, characterise the later stages of D_2. They are commonly associated with shear fracture planes, and normally cut up-section across bedding and S_1 fabrics. In the southern part of Bounds Cliff such thrusts transect downward-facing D_1 folds (Figure 27), and at Barrett's Zawn [SX 027 818] are accompanied by small-scale backthrusts.

The second cleavage is largely absent as a distinct fabric, except in the southern parts of the Bounds Cliff section, where it is first recognised in approximately 0.1 m-thick horizontal shear zones within the Jacket's Point Slate. Farther south, the fault-bounded Harbour Cove Slate shows an intense, steeply south-dipping (60° to 70°) crenulation cleavage, which is axial planar to open, north-verging F_2 folds. These folds have a wavelength of 0.5 m and are seen to refold early D_2 thrusts.

The complex history of D_2 shear is demonstrated by quartz-vein networks associated with the thrust zones. At Jacket's Point [SX 0368 8310], thin quartz veins in the hanging wall of a small D_2 thrust are parallel to the S_1 cleavage, whereas in the footwall they are inclined obliquely to the main fabric. These early veins are seen to be folded by north-verging F_2 folds and both structures are cut across by later vein sets, kinematically related to the D_2 movements. Southwards at Barretts Zawn [SX 0277 8178], tensional vein arrays in association with

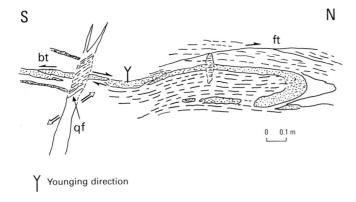

Y   Younging direction

Figure 28   Profile section of an isoclinal south-facing F_1 fold, transected by early D_2 thrusts in the Jacket's Point Slate at Barrett's Zawn [SX 0277 8178]. qf: quartz fibres; ft: forethrust; bt: backthrust.

D_2 structures contain curved quartz fibre growths which record the progressive nature of D_2 shear (Figure 28).

PADSTOW UNIT

The dominant structure within the Padstow Unit is the St Minver Synclinorium (Figure 22) which was recognised by Reid et al. (1910) during the first detailed survey of the region. The axial trace of this regional structure trends east-south-east to west-north-west. It has a wavelength in the order of 6 km, an along strike extention of some 14 km and closes to the east between the Allen Valley and Bodmin Moor. The St Minver Synclinorium has been interpreted as an F_1 structure (Durning, 1989, Warr and Durning, 1988), across which there is a major confrontation in the regional D_1 transport directions (e.g. Andrews, et al., 1988; Durning, 1989). These first-phase folds can be used to divide the Padstow Unit into two structurally distinct subzones: firstly, a northern zone (structures denoted by n), here called the Polzeath Sub-unit, occurs

north of Trebetherick and incorporates the northern limb of the St Minver Synclinorium, and secondly, a southern zone (structures denoted by s), referred to as the Wadebridge Sub-unit, lies to the south in the coastal section of the Camel estuary. The south facing D_1 folds and thrusts of the Polzeath Sub-unit are juxtaposed across the Trebetherick Thrust (Plate 6b) with north-facing D_1 structures of the Wadebridge Sub-unit to give the Padstow Confrontation.

At most inland exposures within the Padstow Unit slaty cleavage dips moderately to gently and shows a highly variable strike orientation across the district. Refolding is probable, and the primary orientation of S_1 is not known. Secondary, non-penetrative cleavages occur at many localities; these show variations in orientation that reflect the sequence of late cleavages observed on the coast.

Polzeath Sub-unit

The Polzeath Sub-unit is limited northwards by high-angle faults and thrusts which juxtapose Trevose Slate against the Bounds Cliff and Tintagel units. To the south it is probably bounded throughout its length by the Trebetherick Thrust. The sub-unit is characterised by south-facing first-phase folds and four overprinting cleavages, each related to a phase of folding. The F_1 folds and S_1 slaty cleavage, and the F_2 and S_2 crenulation cleavage, are important regional structures; the S_3 and S_4 crenulation cleavages are locally developed and associated with only minor folding. The complexities of the geology are seen in coastal exposures cross-cutting the strike of the St Minver Synclinorium from Rumps Point [SW 931 812] in the north to Stepper Point [SW 915 785] and Daymer Bay in the south.

FIRST-PHASE DEFORMATION (D_{1n})

Throughout the coastal section (Figure 29), bedding has been deduced to dip regionally to the north and north-west, but this is largely obscured by intense D_{2n} deformation, so that F_{1n} folds are rarely observed. From Rumps Point southwards to Pentireglaze Haven [SW 933 797] the strata show open to tight, small-scale

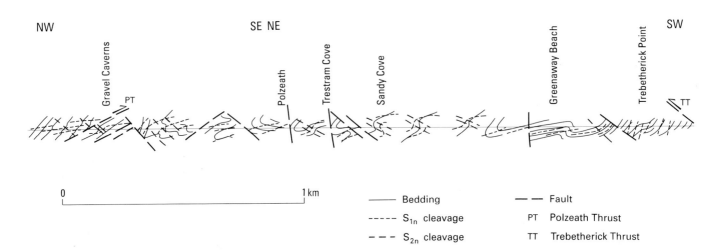

Figure 29   Generalised coastal section through the Polzeath Sub-unit.

(measured in centimetres), north-verging folds which face upwards and to the south. Selwood and Thomas (1988) reported such folds on a metre-scale isocline, refolded in a F_2 antiform at Gravel Caverns [SW 931 798]. A large, northward-closing, F_{1n} fold is also seen in a sea-stack at Trestram Cove [SW 932 790]. It shows a steep north-dipping upper limb, a flat-lying lower limb, and a gently north-dipping slaty cleavage. This fold is believed to have maintained its primary orientation in an area otherwise affected by D_2 folding.

The presence of other F_{1n} folds in the section can be inferred from changes in bedding and S_{1n} cleavage relationships in related localities. A change to south-verging, open folds is thus deduced in, and south of, Greenaway Beach [SW 927 781] (Figure 29) indicating the presence of a large-scale synform. The observed upward south-facing is consistent with an overturned D_1 syncline (Figure 30), the hinge of which is suggested to lie in or near Broadagogue Cove [SW 928 785]. Inland the axial trace of this structure, which formed synchronously with the St Minver Synclinorium, has been identified immediately north of the Trebetherick Thrust. S_{1n} slaty cleavage is always axial planar to large- and small-scale F_{1n} folds; it shows both stylolitic and mica-alignment fabrics. The original attitude of the cleavage is suggested to have been northward dipping at moderate to shallow angles. In the northern sections of Daymer Bay, the bedding and S_{1n} are steeply disposed in asymmetric folding with neutral vergence. This section lies in the footwall of the Trebetherick Thrust and is believed to have suffered rotation effected by the overriding hanging wall.

SECOND-PHASE DEFORMATION (D_{2n})

D_{2n} was a northward-transporting deformation which folded the D_{1n} structures. An early expression of D_{2n} (D_{2na}) is recorded in Trestram Cove (Figure 31), where bedding is displaced to the north along S_{1n} (Figure 31b). This secondary movement along the gently north-dipping S_1 slaty cleavage of a mesoscale F_{1n} fold cuts across thin sandstone beds to form asymmetric augen, indicating northward shear (Figure 31d). The shear bands produced cut across both limbs of the F_{1n} fold. This early northward shear (D_{2na}) was coaxial with the earlier southward transporting D_1 structures. At Trestram Cove, both the S_{1n} cleavage and D_{2na} shears are cut by small (less than 1 m in length) north-dipping thrusts with a southward sense of transport (Figure 31c). Similar thrusts (reorientated by F_{2n} folding) with southward displacements of only a few centimetres are seen at Sandy Cove [SW 929 787] (Figure 31e), where they are close in orientation to S_{1n}. These thrusts are interpreted as the last expression of D_{1n}, and are linked to the development of the much larger Polzeath Thrust at Pentireglaze Haven. This thrust, which carries Gravel Caverns Conglomerate in the hanging wall over Polzeath Slate, is interpreted as a southward-transporting D_{1n} structure (Figure 29). Intense deformation, including ductile and later brittle deformation, is associated with extensive disruption in the hanging wall (Selwood and Thomas, 1988). The present orientation

of the thrust reflects folding about D_{2b} folds, and late faulting.

The main D_{2n} deformation (D_{2nb}) manifests itself as open to close folds of bedding and S_1, which trend north-eastwards with an associated thrusting. These folds, developed on both metre and centimetre scales, show an axial planar, moderate to shallow, south-dipping, crenulation cleavage (S_{2n}), and verge to the north-west. D_{2n} folds are particularly common between Sandy Cove [SW 929 787] and Gravel Caverns [SW 932 797]. Elsewhere, although the S_{2n} cleavage may be well developed, folds are less common. At Trebetherick Point [SW 9245 7792], the steep S_{1n} cleavage is folded into F_{2n} chevron folds by bedding-parallel shear (Figure 32a, Plate 6c). Because the hinge regions coincide with bedding planes, the wavelength of the folds depends upon the original thickness of the beds. Tension gashes in adjacent green slates indicate northward shear (Figure 32b). The second cleavage, which is broadly parallel to bedding and almost at right angles to S_{1n}, can be seen with a scanning electron microscope to comprise anastomosing seams that deform S_{1n} into open crenulations.

Generally S_{2n} is a gently south-dipping, crenulation cleavage, which is axial planar to F_{2n} folds. It is developed throughout the coastal section except at Greenaway Beach, which is positioned close to the core of the regional F_{1n} syncline. S_{2n} is variably developed, appearing as finely spaced black stylolites, as a local, apparently penetrative cleavage, or as a spaced cleavage with planes up to 10 mm apart. In thin section, S_{2n} is always a discrete crenulation cleavage, folding both bedding and S_{1n}. Stylolites are developed within the hinge regions of microfolds, but where intense recrystallisation has taken place, hinges are lost and abrupt planar contacts are found between S_{1n} and S_{2n}.

THIRD-PHASE DEFORMATION (D_{3n})

The D_{3n} deformation generated north-verging similar folds, with a moderate to steeply south-dipping axial planar crenulation cleavage (S_{3n}). The intensity of S_{3n} is highly variable, it may disappear over a distance of a few centimetres.

The D_{3n} folds trend north-east–south-west, with wavelengths of up to 0.3 m (Figure 33), refolding earlier folds and fabrics. At Shingle Gully [SW 9308 7988] F_{1n} folds are deformed around a F_{3n} hinge. Elsewhere, for example at Sandy Cove [SW 929 787] D_{3n} is seen to have developed coaxially with D_{2n} and tightening of F_{2n} folds is observed. Generally, the S_{3n} cleavage dips to the south at a steeper angle than S_{2n}.

Quartz veining along the axial planes of F_{3n} folds at Shingle Gully and Trestram Cove [SW 932790] (Figure 33) possibly developed at the end of D_{3n}. D_{3n} tension gashes at the latter locality are developed on one limb of an F_{3n} fold and these indicate formation by a northward-directed shear.

FOURTH-PHASE DEFORMATION (D_{4n})

Structure and fabrics generated during this deformation show a north-north-west strike contrasting with all earlier fabrics. The deformation is chiefly represented by

Figure 30
Structures at
the facing
confrontation
near Padstow
a. Sketch
geological
map of the
confrontation
zone east of
Padstow.
b. Section A–A′
across the map.
c. Section B–B′
across the map.

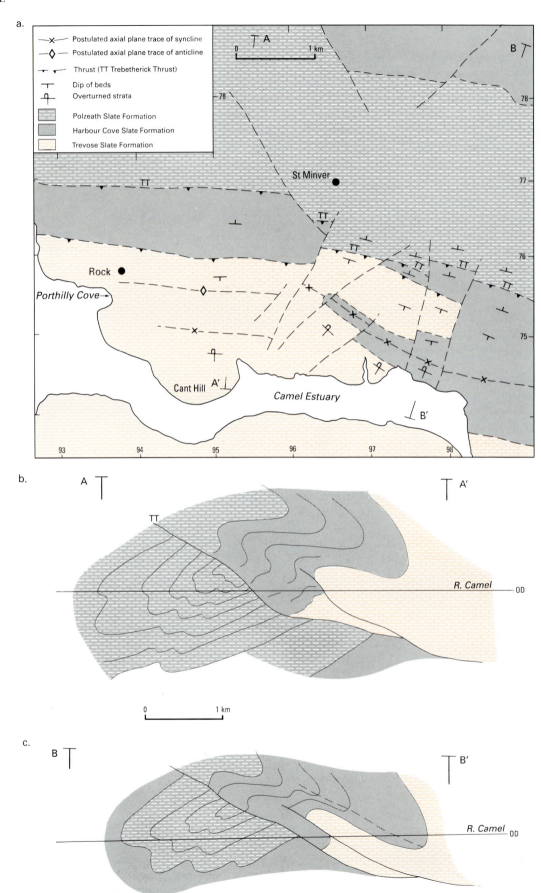

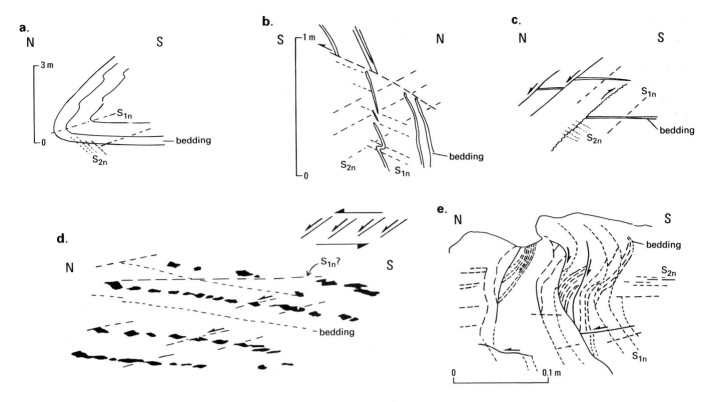

Figure 31 Structures in Polzeath Slate at Trestram Cove (a–d) and Sandy Cove (e).
a. Fold in sea stack viewed from west. b. Upper limb of fold viewed from east, D_{2an} resulted in southward displacement along S_{1n} and was followed by S_{2n}. c. Lower limb of fold illustrated in a. viewed from west, post-D_{2an} thrust (last expression of D_{1n}) with southward displacement, crenulated on S_{2n}. Later northward displacement on S_{1n} also shown.
d. D_{2an} shear bands viewed from west, sandstone beds extended into asymmetric augen by northward shear on S1n. e. Late D_{1n} thrusts, almost parallel to S_{1n}, refolded about F_{2n}.

a steep crenulation cleavage fanning to the west-south-west and east-north-east. It is pervasive in the area between Greenaway Beach and Daymer Bay. Elsewhere it may be absent or only visible as a faint lineation.

The F_4 folds have steep or vertical axial planes trending north-north-west, and fold axes generally plunge towards the north-west at steep angles. Such folds occur in Pentire Haven and are particularly well developed at Greenaway Beach. They appear as monoclines with a wavelength of up to 50 cm, and normally die out over 1 m. These folds are flexural flow folds, formed by east–west compression with simple shear parallel to bedding. Such structures are best developed where the bedding and S_1 cleavage are parallel (Figure 34). Kink bands with the same axial planar orientation are also referred to D_{4n}.

Wadebridge Sub-unit

From the Trebetherick Thrust, the Wadebridge Sub-unit extends southwards beyond the margin of the district and eastwards from the coast to the Cardinham Fault Zone (Figure 22). The area has undergone one main phase of deformation (D_{1s}), followed by three generations of minor folding with each showing an axial planar, crenulation cleavage. Structural relationships are well displayed in cliff sections along the coast and the Camel Estuary. Inland, bedding is difficult to determine in the slates, except at formational contacts and slaty cleavage is the most obvious fabric.

First-phase deformation (D_{1s})

The F_{1s} folds are asymmetric recumbent structures, trending north-east–south-west, and range in wavelength from centimetre to kilometre scales. Mesoscopic folds are tight to close and asymmetric on normal limbs, and gentle to open and asymmetric on steep limbs of the major folds. Fold shape also changes with lithology, being predominantly open to close in the mudstones, but tight in the Staddon Grit and undivided Meadfoot Group. Thin sandstone and limestone beds in thicker slate sequences show intense differential small-scale folding where bedding and cleavage intersect at a high angle. The cleavage relationship to the folds is variable depending upon the location of the fold in relation to higher order structures; thus it may bisect the interlimb angle or approach parallelism with one limb. Hinge zones are commonly rounded, with limestone beds showing hinge thickening; fold axes and bedding/cleavage intersection lineation are parallel, indicating their genetic relationship. In the coastal sections, the

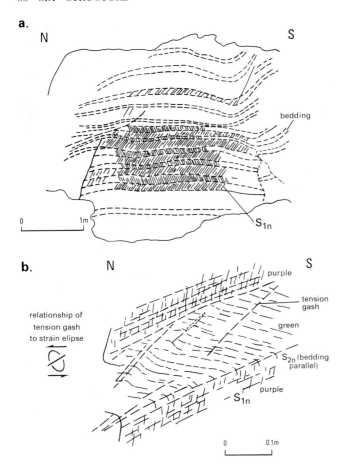

Figure 32 Structures in Polzeath Slate at Trebetherick Point.   a. Cliff section showing horizontal bedding with steep S_{1n} cleavage.   b. Detail showing D_{2n} chevron folding of S_{1n} with tension gashes giving sense of D_{2n} shear.

overall attitude and way-up of the bedding indicates that the sequence lies in the hinge zone and the steep to over-turned northern limb of a regional-scale, asymmetric, northward-verging, antiform. The small-scale folds are buckle folds modified by simple north-north-west oriented shear, which has caused some curvature of fold axes, with flattening parallel to S_l cleavage. Fold vergence indicates the presence of a large-scale recumbent anticlinal fold closure in the area of the Camel Estuary (Figure 35). To the south, folds verge northwards, and show long flat-lying normal limbs and short steep inverted limbs, but north of Padstow, folds verge to the south with long, steep, inverted limbs and short, flat-lying right-way-up limbs. Folds with a southward vergence continue from Padstow northwards to beyond St George's Cove [SW 919 765], where a long, flat-lying fold limb marks a return to north-dipping right- way-up beds with north-verging folds (Figure 35). A synclinal fold closure appears to lie within the area of St George's Cove.

S_{1s} is a finely spaced, penetrative fabric, with no regular pattern of preferred dip direction through much of the coastal section, but south of the outcrop of the Marble Cliff Limestone, the prevailing dip is to the north-east. From

north of Trevone [SW 890 760] to Stepper Point [SW 915 785], cleavage dips gently southwards. Along the shore of the Camel Estuary, S_{1s} dips towards the south-east at about 30°. Bedding/cleavage intersection orientations are highly variable, but the mode is subhorizontal, with an east-north-east trend. Facing along the cleavage is to the north, indicating that the macro-scale structures are upward-facing and shallowly inclined to the south in the northern part of the area, but gently downward facing in the southern part of the coast section. The cleavage is principally a pressure solution phenomenon, but 'chocolate-tablet boudinage' observed in Trevone Bay [SW 8901 7596], where S_{1s} and bedding are almost parallel, suggests that pure shear (flattening) contributed to the formation of S_{1s}. The displacement of bedding along the cleavage, which also gives fold and cleavage mullions in some sections, indicates that northward simple shear was also active at this time. The gentle northward dip of S_{1s} in the south-westernmost parts of the district, which gives downward facing, may be the effect of late regional doming over the buried axis of the granite batholith.

In thin section, S_{1s} stylolites are revealed as anastomosing black seams, axial-planar to microscopic folds, at spacings varying with lithology. Under the scanning electron microscope the stylolites consist of aligned recrystallised micaceous seams separated either by randomly oriented micas or, in the northern part of the area, by aligned micas. In the latter area an early fabric, crenulated or cut across by S_{1s}, has also been observed; this suggests a more intense recrystallisation in the formation of S_{1s} and the transposition of a pre-S_{1s} fabric into the slaty cleavage.

Inland, large-scale folds can be inferred from the repetition of stratigraphical units between the Trebetherick Thrust and the Camel Estuary (Figure 30). The traces of fold axial planes are displaced by cross-faults. Schematic sections across the structure (Figure 30) illustrate that the opposed facing relationships across the Trebetherick Thrust are probably maintained eastwards to the Allen Valley Fault. A facing confrontation has not been recognised to the east of the Allen Valley Fault where steeply

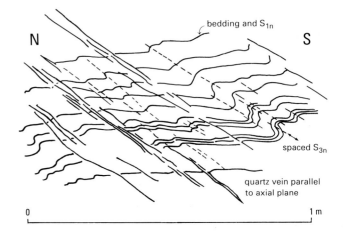

Figure 33 Folds (F_{3n}) in Polzeath Slate at Trestram Cove.

W   E

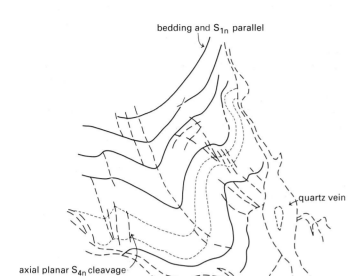

Figure 34 Folding (F_{4n}) in Polzeath Slate at Greenaway Beach.

dipping strata mapped within the closure of the St Minver Synclinorium are cut by west to south-west gently dipping slaty cleavage, associated with minor north-verging folds.

East of Wadebridge, structures are locally reorientated parallel to an arcuate topographic lineament, lying within the outcrop of the Trevose Slate, which can be traced from north of Spittal [SX 0780 7310] south-westwards towards Sladesbridge [SX 0106 7146]. The reorientation effects extend southwards for up to 1.5 km. Farther south, small-scale structures appear which are interpreted to lie in the steep limb of a major north-verging fold. The shortening effected by this early ductile folding event was increased by later minor thrusting, cutting up-sequence towards the north west, and by associated minor asymmetrical folding. Such structures are comparable to those observed within the Wadebridge Sub-unit between Mawgan Porth [SW 848 672] and Trevose Head.

D_{1s} THRUSTS

The northern boundary of the main outcrop of the Staddon Grit dips south at a low angle. Small outliers at Trethewell [SW 869 702], Porton Down [SW 954 690] and Haycrock [SW 970 694] appear to have horizontal bases. The boundary is exposed in the cliffs southward from Carnewas Point [SW 8475 6895] where it is a south-dipping planar fault. South of Pendarves Point [SW 8485 6930], thinly bedded and interlaminated fine-grained sandstone and siltstone beds in the Bedruthan Formation occur in open recumbent mid-mesoscopic scale folds about gently northward-dipping S_{1s}. Southwards towards the fault, folds in the Bedruthan Formation are tighter,

and S_{1s} cleavage dips to the south, parallel to the fault. In the hanging wall, folds in the thicker sandstones of the Staddon Grit are recumbent, and vary in amplitude and tightness with some almost isoclinal in style. Bedding on either side of the fault is disrupted, commonly parallel to S_{1s} which is not itself deformed significantly by later structures. Limb lengths are such that, to either side of the fault, the fold envelopes are moderately to steeply inclined and overturned northwards. The boundary is interpreted as a D_{1s} thrust which transports Staddon Grit northwards over the Bedruthan Formation.

Similar early thrusting may also determine the local distribution of the Bedruthan Formation in the vicinity of Burlawn [SW 996 702]. There the Bedruthan Formation caps a flat-topped hill with Trevose Slate on the flanks. The Bedruthan Formation is mostly moderately to steeply inclined in that area but, about 400 m south of Burlawn [SW 999 698], the Bedruthan Formation—Trevose Slate boundary shows a change from a steep to a gentle dip; thrusting rather than large-scale recumbent folding is considered the more likely explanation. At Mawgan Porth [SW 8487 6769], the Meadfoot Group rocks are interpreted as being thrust northwards over the Staddon Grit. Although showing small-scale folds, in general the underlying rocks are steeply overturned. The Meadfoot Group rocks appear to have low dips and to be the right way up. The thrust may have developed in the hinge zone of the major antiform; it has emplaced rocks from the gently dipping southern limb onto the steeply dipping northern limb.

POST-D_{1s} DEFORMATION

Post-D_1 deformation is seen as local, open and tight folds of slaty cleavage on a centimetre and metre scale, with crenulation cleavages developed axial planar to the folds. Thrusts and kink bands are also present. Most of these structures are minor, and rarely persist for more than a few metres. No major refolding of earlier structures has been recorded. Three generations of crenulation cleavage have been identified with differing strikes: north-east–south-west (S_{2as}), west-north-west–east-south-east (S_{2bs}) and north-north-west–south-south-east (S_{3s}).

Three sets of S_{2as} cleavages have been identified: a set of steep cleavages dipping 80° to 90° N or S and a conjugate pair less than 60°N and 25° to less than 60° S with an interlimb angle of more than 60°. The south-dipping member of the conjugate pair is almost coaxial with S_{1s}, and associated with northward-verging open buckle folds of bedding and S_{1s}. In local zones about 1 m wide in the coast section, this S_{2as} cleavage is closely spaced, and transposes S_{1s} to become the predominant cleavage.

These later structures suggest a continuation of the D_{1s} event, but since the morphologies are quite distinct from F_{1s} and S_{1s}, the deformation has been assigned to D_{2s}. The north-dipping S_{2as} crenulation cleavages occur sporadically in the cliff sections south from Trevose Head, either steeply or gently inclined, and associated folds are variable in scale in the mesoscopic range. Minor folds are open to close, gently to moderately inclined,

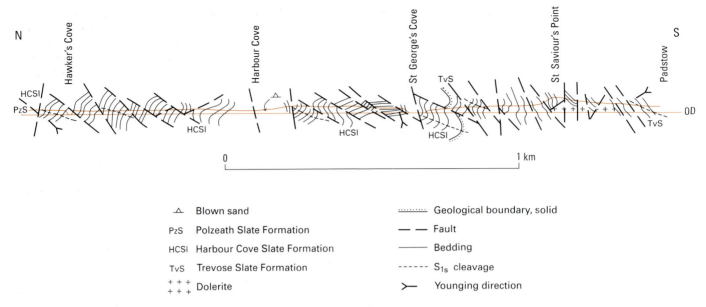

	Blown sand		Geological boundary, solid
PzS	Polzeath Slate Formation	— —	Fault
HCSl	Harbour Cove Slate Formation	———	Bedding
TvS	Trevose Slate Formation	------	S_{1s} cleavage
+ + + + + +	Dolerite	⋟	Younging direction

Figure 35 Generalised coastal section through the Wadebridge Sub-unit in the Camel estuary.

subhorizontal, verging southwards. In the cliffs between Pentire Steps [SW 848 704] and Bedruthan Steps [SW 849 694] larger mesoscopic monoform folds of S_1 verge to the south. Their shorter, steeply inclined limbs with a length of 15 m or more, comprise metre-scale, parasitic folds cascading downwards to the south, with cleavage well developed only in the fold hinges. In the Falmouth district, Leveridge et al. (1990) have identified two similarly oriented phases of deformation, of which the southward-vergent phase preceeds the northward-vergent phase. However, Turner (1968) proposed that they form a conjugate system in the area immediately to the south of the Trevose Head and Camelford district, and his interpretation is followed here, although no new information is available. Folds associated with S_{2bs} are observed to refold S_{2as} structures in Kestle Quarry [SX 015 715]. The S_{2bs} cleavage is very locally developed and dips to the north or south; it indicates a change in compression direction during D_{2s} from north-west–south-east, to approximately north–south.

THRUST FAULTS (D_{2s} AND D_{3s})

In the cliffs south of Trevose Head, flat-lying thrust faults associated with D_{2s}, vary from mid mesoscopic scale dislocations related to isolated fold couplets, to much more extensive structures traceable over several hundreds of metres, as for example at Wincove Point [SW 854 738]. Commonly S_{2s} and F_{2s} are present in the lowest metre of the hanging wall and are also present in small duplex systems.

The Trebetherick Thrust is mapped some 5 km eastwards from the Camel Estuary towards the Allen Valley Fault. Associated structures indicate northward transport on the fault in late D_{2s} times, and that it was coeval with the formation of the S_{2s} cleavage. Although in the coastal section this thrust delimits the facing confrontation

dividing the Wadebridge and Polzeath structural sub-units, it has not been mapped through the closure of the St Minver Synclinorium. This suggests that there was no major displacement along the thrust.

In the Camel Estuary sections, minor thrust faults are recorded striking 060° to 080°, with dips of about 40° southwards, or with shallower dips averaging 18° south. Measurement of shear fibres on surfaces indicate dip-slip movement. Thrust structures associated with northward transport of the hanging wall are revealed in a lane cutting at Trevelver Farm [SW 959 751] (Figure 36). In the Atlantic coastal section, minor faults showing southward back-thrust displacement are closely associated with some of the isolated minor folds and the larger-scale structures. There are also isolated duplex systems in the Trevose Slate with floor and roof thrusts up to 2 m apart; minor folds and cleavage of D_{2s} style and orientation are associated with the intervening imbricate thrusts. These faults formed with a north–south compression and are a brittle

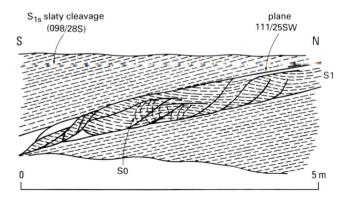

Figure 36 Thrusting in the Trevose Slate, exposed on the west side of lane leading to Trevelver Farm [SW 959 751].

expression of D_{2s}, which later produced the dextral north-west-trending strike-slip faults which cut across the thrusts.

The S_{3s} cleavage is axial planar to tight, upright to inclined folds, usually with limb lengths of less than 2 m, although rare examples with short limb lengths of up to 6 m occur on the coast. Most folds have gently plunging, rounded hinges, and the inclined folds verge to the east-north-east. Usually the cleavage dips vertically or steeply to the north-east or south-west, but at some localities cleavage dip steepens upwards. Its relationship to other crenulation cleavages is not observed, but its widespread distribution suggests that it, and associated kink bands, were the last to develop, forming within a north-east–south-west compressive regime (D_{3s}).

The gently inclined thrust faults which are associated with most occurrences of gently inclined F_{3s} folds, range from single, simple dislocations of limited extent to more complex imbricate and duplex systems, traceable over several hundreds of metres in coastal sections. Contraction kink bands are developed parallel to the strike of all three crenulation cleavages, but those trending north-north-west are dominant. The dips of all axial planes are vertical or steeply inclined. A north–south weakly developed, crenulation cleavage associated with gentle warping of earlier planar structures including D_{3s} thrusts, is present locally in the southern coastal sections, for example at Booby's Bay [SW 8573 7547]. There, low-angle faulting associated with gently inclined, tight folds verging westwards is also sporadically developed. An associated cleavage is gently inclined eastwards. A weak cleavage with similar orientation is seen to deform S_{3s} elsewhere, as at Pentire Steps [SW 8485 7025].

Correlation of deformation episodes of the Padstow Unit

The principal structural difference between the Polzeath and Wadebridge sub-units is the fold-facing direction. In the Wadebridge Sub-unit, the first deformation (D_{1s}) gave rise to the north-facing folds (F_{1s}), whilst in the Polzeath Sub-unit the first deformation (D_{1n}) gave south-facing folds during southward transport. Although D_{1s} and D_{1n} are distinct, subsequent minor overprinting deformations can be matched (Figure 37).

The effects of east–west compression, which gave rise to local buckling and flexural folding, and to the development of a north-north-west-trending, steeply dipping, crenulation cleavage, is present in both sub-units. In the Polzeath Sub-unit, due to its more complex early history, this is the fourth deformation event (D_{4n}), but in the Wadebridge Sub-unit it is only the third (D_{3s}) (Table 9).

The development of a north-east-trending, moderate or steeply south-east-dipping, crenulation cleavage, which is axial planar to folds of pre-existing fabrics, is also seen in both sub-units. Within the Polzeath Sub-unit, examples of these structures are abundant, and can be identified as the third deformation (D_{3n}) to affect the rocks, folding D_{1n} and D_{2n} structures. Within the Wadebridge Sub-unit examples of such folds are scarce but examples of the accompanying cleavage (S_{2as}) are abundant, and identified as being generated during the second deformation (D_{2s}). The two units are juxtaposed by the Trebetherick Thrust, one of a series of north-transporting, east–west-trending thrusts which have been assigned to a D_{2s}/D_{3n} event.

Prior to the D_{2s}/D_{3n} event, the Wadebridge Sub-unit exhibits only one deformation (D_{1s}), whilst the Polzeath Sub-unit exhibits two (D_{1n} and D_{2n}). Within the Polzeath Sub-unit the D_{2n} event gave rise to asymmetric north-verging folds with north-east- or south-west-plunging fold axes, and a shallow southerly dipping cleavage. The attitudes of folds and cleavage are identical to the D_{1s} within the Wadebridge Sub-unit, where F_{1s} folds have axes gently plunging north-east or south-west, and axial

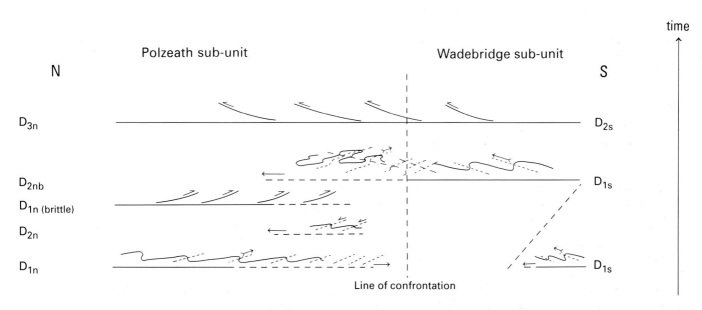

Figure 37   The sequential development of structures across the Padstow Confrontation.

Table 9 Summary of deformation events in the structural units.

Padstow Unit		Bounds Cliff Unit	Tintagel Unit
Wadebridge Sub-unit	Polzeath Sub-unit		
D_{3s}: NE–SW compression Local folds verge ENE steep S_{3s} crenulation cleavage and kink bands	D_{4n}: E–W compression Local NNW monoclinal folds and kink bands; sporadic S_{4n} crenulation cleavage		D_3: extensional faulting Late D_2: brittle thrusting
D_{2s}: N–S compression Local monoclinal open folds, local variable S_{2s} crenulation cleavage some thrusting	D_{3n}: Northward shear Local N-verging open folds, with S-dipping axial planar S_{3n} crenulation cleavage	D_2: NNW shear Local folding with S_2 crenulation cleavage NNW-transporting thrusts	Early D_2: ductile deformation Large-scale NNW transport; shear folds, thrusts, stretching, reworking of earlier cleavage. M_2 greenschist metamorphism
D_{1s}: folding N-facing tight folds with S_{1s}, NNW-transporting thrusts	D_{2n}: Northward shear Variably developed N-verging folding and thrusting, gently S- or N-dipping S_{2n} crenulation cleavage		
	D_{1n}: folding S-facing folds with S_{1n}, S-transporting thrusts	D_1: folding S-facing folds with S_1 cleavage	D_1: folding tight-isoclinal S-facing folds with S_1, and local S-transporting thrusts M_1 metamorphism

Terminology — D_1, D_2 etc. relate to relative age within, but not correlation between, structural units/sub-units

planar S_{1s} cleavage dipping shallowly to the south. This similarity was used by Durning (1989b) to suggest that the D_{1s} event which affected undeformed rocks in the Wadebridge Sub-unit, is the same event which gave the D_{2n} structures in the Polzeath Sub-unit, which contained a pre-existing deformation event (D_{1n}).

The D_{1n} event of southward transport has no apparent equivalent within the Wadebridge Sub-unit. It appears that the effects of the D_{1n} deformation died out southwards as the effects of the strain decreased. It is significant that observations of bedding and S_{1n} cleavage in the Polzeath Sub-unit show them to be parallel (for example Greenaway Beach) or at a close angle. This is similar to the pre-S_{1s} fabric (p.67) in the Wadebridge Sub-unit. It is suggested that the pre-S_{1s} fabric was developed in an area close to the minimum development of the S_{1n}, and that farther south, the cleavage was too weakly developed to be preserved. The northward transition into the stronger (S_{1n}) fabric is lost by shortening across the Trebetherick Thrust which now forms the confrontation, so that S_{1n} appears abruptly. The correlation of the pre-S_{1s} fabric with S_{1n} is speculative, but its restricted occurrence in the Wadebridge Sub-unit argues against it being a syn-sedimentary or early diagenetic feature. A schematic section across the Padstow Confrontation employing the correlation devised above is illustrated in (Figure 38).

The south-west-dipping cleavage recognised east of the Allen Valley Fault correlates most readily with the S_{2n}/S_{1s} cleavage represented in the Polzeath and Wadebridge sub-units respectively, but as a slaty cleavage, the more direct correlation is with the S_1 cleavage developed in the Wadebridge Sub-unit. As such, it would postdate the regional D_{1n} south-facing regional syncline developed along strike in the Polzeath Sub-unit.

The above correlations indicate that deformation began in the north with a southward transporting event (D_{1n}), which generated kilometre-scale, south-facing recumbent folds. This ductile event died out southwards, where it met an early expression of northward-transporting deformation (D_{2n}), which produced local shearing along the S_{1n} cleavage, and shear bands cutting across pre-existing F_{1n} folds. D_{1n} continued as a brittle event, producing small and large-scale thrusts (for example the Polzeath Thrust) which cut across the D_{1n} ductile and D_{2an} structures.

The main northward-transporting deformation (D_{1s}) followed these events, and overprinted the earlier structures to produce $D2_{bn}$ structures. South of the southern limit of D_{1n}, the northward-transporting deformation was the first to affect the rocks (D_{1s}), and produced north-facing F_{1s} folds and S_{1s} slaty cleavage. Continued northward-transporting deformation imposed a further

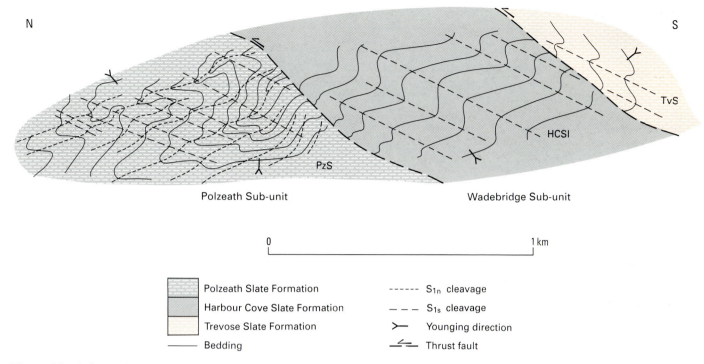

Figure 38 Schematic section through the Padstow Confrontation at Rock [SW 940 760].

ductile folding event on both the Polzeath (D_{3n}) and Wadebridge (D_{2s}) sub-units. This was followed by a thrusting event which shortened the structure, and brought rocks with first phase (D_{1s}) north-facing structures (Wadebridge Sub-unit), over rocks to the north (Polzeath Sub-unit) which have first phase, south-facing structures (D_{1n}). The confrontation of facing directions was thus established. Finally a later change in compression direction generated another minor fold phase (D_{3s}/D_{4n}).

Cardinham Fault Zone

The Cardinham Fault Zone constitutes a structurally complex area 3 km wide, extending north-north-west to south-south-east at the western margin of the Bodmin Moor granite (Figure 39). The zone is largely developed within uniform Trevose Slate Formation and is poorly exposed.

The complex network of valleys developed by the Fowey and Camel river systems suggests that the fault complex is an anastomosing system with individual faults seldom more than 2 km long. The fault pattern suggests the development of shears, thrusts, and normal faults in a major dextral wrench system. Detailed stratigraphical and structural analyses have not been possible, but complexities comparable to those described by Turner (1984) in the Tamar Fault Zone farther east are possibly present.

Where unmodified by later movements, slaty cleavage in the fault zone dips south-west. Evidence for associated folding is limited to fault-controlled valleys lying at the western margin of the fault zone west of Cardinham. These reveal anticlinal cores of Bedruthen Formation in Trevose Slate, aligned north-west to south-east, which may have been formed by transpression in the dextral system.

The fault zone lies in the metamorphic aureole of the granite along much of its length, and probably controlled the south-western margin of the granite. The modification of pre-granite structures does not extend more than 1 km from the granite boundary. The uplift of the country rocks, associated with the granite intrusion, may have been accommodated by movements along fractures within the fault zone.

The Cardinham Fault Zone forms a segment of the St Teath–Portnadler Fault Zone, proposed by Dearman (1963) as one of a series of major north-north-west-trending dextral faults which traverse the Cornubian peninsula; he interpreted these faults as Armorican tensional structures rejuvenated during the Alpine orogeny. Recent mapping has shown St Teath to lie outside the main structure; the Cardinham–Portnadler Fault Zone is therefore a more appropriate name.

The diverse structures recognised within this fault zone are the expression in the cover of movements on a deep fracture, which played a critical role in the origin and development of the Trevone Basin (Selwood, 1990). Facies changes across the zone signal the early activity of the fracture which also provided the pathway for the intrusion of dolerite dykes. Later it was to compartmentalise the regional structure, allowing separate development of the Trevone Basin and the Liskeard Rise.

Alternative views on the structure of the Padstow Unit

The alternative views of Selwood and Thomas on the structure of the area of the Padstow Unit have been presented recently in Selwood et al. (1993). Rather than the Padstow Unit comprising a complexly deformed

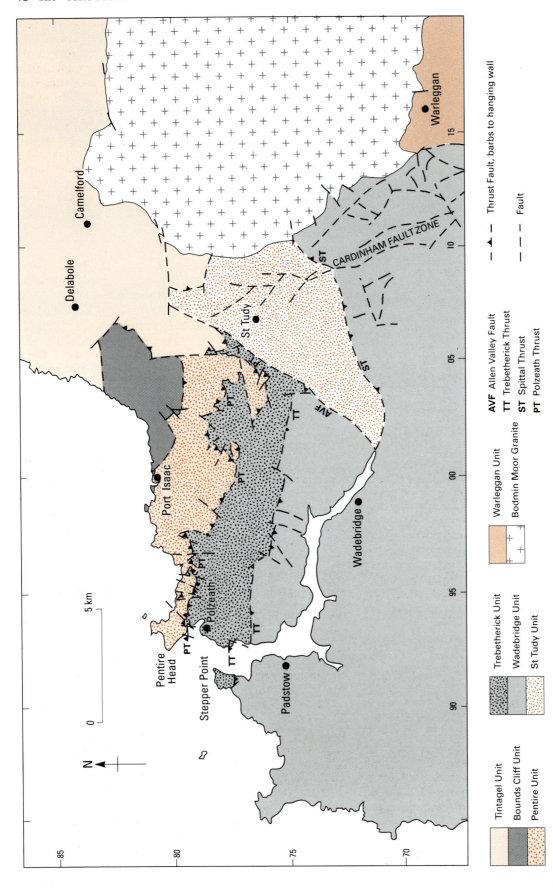

Figure 39 Alternative interpretation of structural units by Selwood and Thomas (1993), showing major fault systems.

synform in a simple conformable basinal succession, as presented above, they propose subdividing the area into four structural units, the Pentire, Trebetherick, Wadebridge and St Tudy units (Figures 39, 40). These are based on their new mapping and biostratigraphical interpretations. Their Pentire and Trevone successions are juxtaposed at the Polzeath Thrust (D_{1n}), and a St Tudy structural unit is differentiated east of the Allen Valley Fault (Figure 39). Within the area, mapped as the closure of the St Minver Synclinorium on the 1:50 000 map, their St Mabyn Succession (Selwood et al. 1993) (Figure 40) youngs eastwards.

In the St Tudy Unit, formational contacts are reported to dip steeply westwards to produce the arcuate outcrop pattern of neutral folds shown on the map. These beds are cut by a slaty cleavage dipping uniformly at moderate angles towards west-south-west or south-west. This cleavage which maintains its attitude unaltered towards the granite, is seen to be axial planar to small-scale local folds. The steep orientation of strata appears to be a pre-slaty cleavage structure.

The southern limit of the unit is stratigraphically defined (Figure 39), and corresponds to an arcuate topographic lineament which can be traced from north of Spittal [SX 0780 7310] south-westwards towards Slades-bridge [SX 0106 7146]. Across this line there is a change in structural style. The line is interpreted as a D_{1s} thrust (Spittal Thrust) carrying the Wadebridge Unit over the St Tudy Unit. Within the Wadebridge Unit immediately to the south, thin lenticular tuffs, distinctive pelitic horizons and the strike of associated south-dipping cleavage, are all arcuately disposed broadly parallel to this feature. But through a zone extending south for 1 to 1.5 km, this structure is lost, and is replaced by tight to isoclinal, north-facing folds, which trend regularly east-south-east.

Since the Pentire and Trevone successions appear in normal, gently northward-inclined sequences within the area differentiated as the northern limb of the St Minver Synclinorium on the 1:50 000 map, and the younging of strata in its supposed closure is to the east rather than west; the existance of this regional structure appears highly improbable.

WARLEGGAN UNIT

The Warleggan Unit crops out south of Bodmin Moor and east of the Cardinham Fault Zone (Figure 22). D_1 structures within it are revealed by beds of calcflintas in the Rosenun Slate. These beds appear in the cores of doubly plunging, tight, angular anticlines with vertical axial surfaces. Hinge lines are oriented east–west, and individual folds can be traced for up to 2 km in the vicinity of Panters-bridge [SX 157 679]. Some reorientation of D_1 fold culminations indicated by calcflinta pods, has, however, taken place west and south-west of Mount [SX 147 680] where the beds are affected by the Cardinham Fault Zone.

The slaty cleavage (S_1) and cleavage-bedding intersection lineation of the D_1 deformation have been wholly obscured by the transposition of a gently south-dipping S_2 cleavage. D_2 folds have not been identified, but a transi-

tion from ductile to brittle deformation in D_2 is indicated by minor thrusting cutting steeply up-section towards the north. The northern boundary of the Pantersbridge calc-flinta body [SX 1580 6862] is a thrust contact; fine-grained fault rocks and silicified breccias are exposed in the Warleggan River section [SX 158 682]. Within the same body, silicified lenses of fault rock, and low amplitude folds of S_2 occur in the hanging wall of a thrust, exposed in a road cutting [SX 1576 6826] west of Pantersbridge. These distinctive structures represent a northward continuation of structures identified by Burton and Tanner (1986) on the Liskeard High (Selwood, 1990) separating the Trevone and South Devon sedimentary basins.

LATE FAULTS

Steep to vertical faults which cut across the main phase Variscan structures are extensively developed throughout the district. Regionally they can be grouped into four main trends: east-north-east–west-south-west, north–south, and a conjugate set of north-west-trending and north-east-trending strike-slip faults. In addition, there are rare north-west-trending thrust faults and north-east-trending normal faults.

East-north-east–west-south-west faults

East-north-east-trending and east-west-trending faults are prominent in the district, and commonly mark formational boundaries. The faults are vertical to steeply inclined and may show normal, reverse and strike-slip movements. The complex history of movement deduced for this set of faults suggests frequent reactivation of persistent fractures. For example, on Trevose Head [SW 8678 7602], moderately to steeply inclined south-south-east-dipping faults show reverse displacements, but later minor down-dip movements. These faults have a similar orientation to the major faults which appear to have controlled the development of the basin. A series of east-north-east-trending faults cut the margin of the Bodmin Moor granite at Devil's Jump [SX 102 804], and show dextral displacements of up to 220 m. The most continuous of these can be traced for 10 km; it is refracted into a north-east–south-west trend through the north-western tip of the batholith, and emerges at Lanlavery Rock [SX 1562 8261] with east–west strike parallel to the northern margin of the granite. The north-western spur of the Bodmin Moor granite is downfaulted towards the north-west along this fault line. Mineralisation has occurred along these lines at Tregardock Mine [SX 0447 8404] and Roughtor Mine [SX 1630 8270].

At Carwen [SX 1085 7381] the faults cutting the granite are sinistral, show oblique-slip movement, and dip steeply towards the south. In the Cardinham area, comparable structures display an approximately east–west orientation; normal dip-slip displacements throw southwards on steep fault planes which have undergone extensive haematisation. The coincidence of east–west faults with the strike of pervasive haemati-

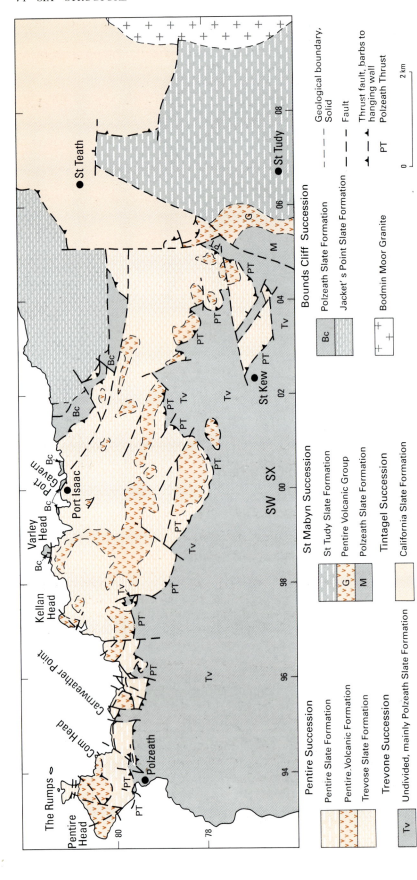

Figure 40   Alternative interpretation of the geology of the area between Pentire Head and St Teath (Selwood and Thomas, 1993).

sation, quartz-porphyries and tin-tungsten-copper mineralisation suggests that they were all emplaced during the same stress episode that produced the east–west extensional faults.

Erosion along a set of discontinuous east-north-east-trending faults plays a major role in controlling coastal morphology between Rumps Point [SW 931 812] and Tresungers Point [SX 009 811]. Reverse movements on such faults, which dip steeply southwards, carry the Trevose Slate over Polzeath Slate and Harbour Cove Slate in the headlands of Varley Head [SW 9844 8134], Lobber Point [SW 9926 8117], Port Isaac [SW 9976 8115] and Tresungers Point. Continuity along strike is interrupted by late north-west-trending normal faults which downthrow to the north-east. The boundaries of the Harbour Cove Slate, in the northern limb of the St Minver Synclinorium some 2 km to the south, are mainly vertical and are likely to be defined by east-north-east-trending faults.

North–south faults

The north–south faults in the district are vertical and can be traced over long distances as linear topographical features. Where they intersect the coast, they show broad fault-zones with structures indicating large vertical movements; quartz fibre growths suggest a component of dextral slip. Many of the south-south-east-trending faults on the coast south of Trevose Head are moderately inclined normal faults, with a few metres throw. Other moderately inclined faults, and the steeply inclined faults show dextral, strike-slip movements. At Mother Ivey's Bay a fault displaces dolerite sills by 750 m, from Merope Rocks [SW 862 766] to Little Cove [SW 865 760]. Faults displace the boundary between the Bedruthan Formation and the Trevose Slate south of Porthcothan [e.g. SW 854 701].

Mineral lodes have been worked in north–south oriented vein systems (crosscourses) throughout the district (see Chapter Seven). Faults following the same trend are developed north-west of the Allen Valley Fault, where they result in large dextral displacements of formations in the northern limb of the St Minver Synclinorium. For example, the Harbour Cove Slate shows a dextral displacement of more than 1 km by a fault which passes through Trequite [SX 0288 7694], and a parallel fault 0.8 km to the east offsets the boundaries by more than 0.5 km. Traced northwards, both these faults show a reduction in strike-slip movement where they cut across structures developed in the Bounds Cliff and Tintagel successions. Such changes along strike may indicate that a complex movement history postdates the east–west faulting and predates the final emplacement of the granite batholith. In addition, north–south faults offset the southern margin of the granite and Exley (1965) recorded fault zones of this orientation within the granite.

North–south faults are rare in the Wadebridge Sub-unit; they trend between 165° and 200°, and dip averages 20° westwards. These dip-slip faults developed from east–west extension, and may have formed in response to the relaxation of stress which formed the last (S_3) crenulation cleavage. Some of the north–south faults recorded

in the coastal section [SW 8572 7447] south of Trevose Head dip gently eastwards, and are associated with westward verging thrust movements. The steeply dipping faults have a normal sense of movement.

North-west–south-east and north-east–south-west strike-slip faults

This conjugate set of faults forms prominent topographical lineaments which are expressed as major gullies in the coastal section. The faults are vertical, and most readily identified inland where they produce strike-slip displacements of formational boundaries and of east-north-east-trending and north–south-trending faults.

The north-west-trending Cardinham Fault Zone is the principal dextral, strike-slip structure in the district. Medium-scale, strike-slip faults developed across the district are thought to be associated with late movements on this structure. At the southern margin of the St Minver Synclinorium, north-east-trending dextral strike-slip faults displace the Trebetherick Thrust. Both sets of faults are linked to late north-south compression in D_{2s}. Initially, this produced thrusts cutting up-structure, and was followed by strike-slip movement. Shear fibres on fault planes indicate some vertical movement, possibly as late-stage adjustment. East-south-east-trending faults, in the coast section [SW 8495 7159] south of Trevose Head, that dip moderately to steeply south-south-west show normal displacements, but slickensides also record dextral strike-slip movement. A lead- and silver-bearing lode has been mined in sinistral strike-slip structures at Pengenna [SX 0487 7882] and Treburgett [SX 0574 7982].

North-west–south-east thrust faults

In the Wadebridge Sub-unit rare thrust faults with low-angle westerly dips deform slaty cleavage in their hanging walls to form east-verging folds which may show an axial planar crenulation cleavage (S_{3s}). These thrusts appear to have formed in a north-east–south-west compression, possibly related to the development of the S_{3s} crenulation cleavage, and associated kink bands (D_{3s}).

North-east–south-west normal faults

Late, dip-slip, extensional faults with north-east–south-west trend are common in the Wadebridge Sub-unit. Observed fault planes dip shallowly to the south-east or, more commonly, steeply northwards. The latter faults, which are seen to displace formational boundaries inland, may represent reactivated conjugates to D_{2s} north-west–south-east-trending dextral strike-slip faults.

REGIONAL METAMORPHISM

Tintagel High Strain Zone

Petrological investigations in north Cornwall have focused on the complex character and development of regional metamorphism in the Tintagel High Strain Zone

(Sanderson 1979). A complete section through the zone is exposed in coastal sections between Tregardock Beach [SX 040 850] to the Rusey Fault [SX 125 935] in the Tintagel district. The southern boundary of the zone has not been delimited, but metamorphic intensity appears to decrease gradually southwards across the Tintagel Unit in this district. The pressure-temperature (P-T) history in the High Strain Zone has been documented by Primmer (1985 a, b), who recognised a composite early event (M_1) in which mineral assemblages indicating upper anchizone peak temperatures of 250 to 300°C were overprinted by a greenschist-facies metamorphism. This was followed by a second event (M_2) at higher temperature (up to about 500°C), but lower pressure. The final phase was a retrograde metamorphism at temperatures of 200 to 300°C. This sequence of events is characteristic of thrust-related metamorphism, and is consistent with rapid tectonic burial followed by thermal re-equilibration and subsequent exhumation.

Correlation of these events with the deformation chronology is complicated by the fact that although the majority of the high strain is associated with D_2, this is broadly coaxial with D_1 (Andrews et al., 1988). Pamplin (1990) has suggested that the M_1 event is associated with D_1 and the development of the S_1 fabric (Table 9). The slaty cleavage in the High Strain Zone appears as a penetrative muscovite-chlorite foliation in the slates and as a chlorite-dominated foliation in the volcanic rocks. This fabric was modified by later greenschist-facies metamorphism. The occurrence of pyrophyllite in slate of the Delabole Member lying outside the High Strain Zone, suggests peak temperatures of 250 to 300°C during the M_1 crystallisation of the D_1 slaty cleavage (Primmer, 1985a). This is consistent with the upper anchizone metamorphism of M_1 elsewhere in the district.

Following the crystallisation of this M_1 fabric in the High Strain Zone, an early phase of porphyroblast growth was initiated under the same regional stress (Primmer, 1985a). Chloritoid or chlorite were sporadically developed in the Tredorn Slate Formation, and the textures of Mn-rich garnets in the Barras Nose Formation suggest rapid growth under prograde metamorphic conditions. Penecontemporaneously, porphyroblasts of unzoned Fe-rich epidote formed in the volcanic rocks; these indicate greenschist-facies temperatures (about 400°C). This late M_1 event of Primmer (1985a) correlates with the high-strain episode of D_2. The greenschist-facies metamorphism of D_2 extends into the later M_2 of Primmer (1985a). This is indicated by the onset of a second phase of static growth of inclusion-free, randomly orientated porphyroblasts of biotite or actinolite, which indicate a relaxation of regional stress, but raised temperatures (up to 500°C), immediately following D_{2b}.

Within the Delabole Slate much of the chloritoid became rimmed with platy muscovite, and the garnet-bearing slates show a development of biotite laths crosscutting garnet rims and the foliation. Biotite or actinolite porphyroblasts were developed in the volcanic rocks, where they replace chlorite along S_1. The M_2 porphyroblasts are aligned in the slaty cleavage, and define the north-north-west–south-south-east mineral lineation which is synchronous with the development of the Davidstow Anticline (D_2).

Retrogressive metamorphism in the slates resulted in progressive pseudomorphous replacement of chloritoid by fine-grained chlorite and muscovite aggregates. This suggests that the stress declined but the temperature remained relatively high (200° to 300°C).

Away from the Tintagel High Strain zone, slates in the area of anchizone metamorphism in the district (Figure 41), are characterised by assemblages of quartz-albite-white mica-chlorite-calcite-dolomite ± pyrophyllite with iron (titanium) oxides and sulphides in accessory amounts. In this area, Warr et al. (1991) have described chlorite/paragonite stacks which formed before S_{1n} cleavage. The origin of these porphyroblasts is uncertain although they may have formed from detrital biotite, but it is clear that recrystallisation occurred before the onset of the first regional deformation. The persistance of pyrophyllite suggests that greenschist grades were not attained (Primmer 1985a).

Metabasic rocks

The basaltic rocks, in some cases, proved more sensitive than the pelites to mineral change at the low temperatures and pressures involved in the anchizone metamorphism. However, most pillow lavas probably underwent submarine alteration following eruption and were subsequently altered during regional metamorphism to non-diagnostic chlorite-albite-carbonate-sphene-iron oxide assemblages that replace all primary material. Metabasites that formed high-level intrusions co-genetic with an extrusive lava/volcaniclastic suite (Rice-Birchall and Floyd, 1988) are likely to have been affected by hydrothermal circulation under an enhanced geothermal gradient. In consequence, their observed secondary alteration is likely to be the product of both the hydrothermal and the low-grade metamorphism. The patchy replacement of primary phases in dolerite intrusions, gives rise to diagnostic assemblages of albite-prehnite-pumpellyite and chlorite-epidote-prehnite-actinolite in internal alteration domains, and non-diagnostic chlorite-albite-carbonate in the finer-grained marginal facies. All the mafic phases in the hydrous dolerites are variably replaced by chlorite, that ranges from diabantite and pychnochlorite in the St Saviours sill [SW 9224 7601] and to brunsvigite at Merope Rocks [SW 861 766], Trevone Bay (Floyd and Rowbotham, 1982; Rice-Birchall, 1991). This suggests that chlorite compositions reflect the composition of the bulk rock, rather than that of specific alteration sites or the metamorphic grade (cf. Cho and Liou, 1987). Actinolite replaces both kaersutite and chlorite in these rocks, and appears to be generally later in development than the other secondary phases. In the anhydrous dolerites, actinolite replaces clinopyroxene and secondary chlorite.

At Stepper Point [SW 915 784], a single phase (plagioclase) is replaced by the equilibrium assemblage prehnite-pumpellyite-epidote-albite. This is characteristic of the prehnite-pumpellyite facies, within the petrogenic grid of

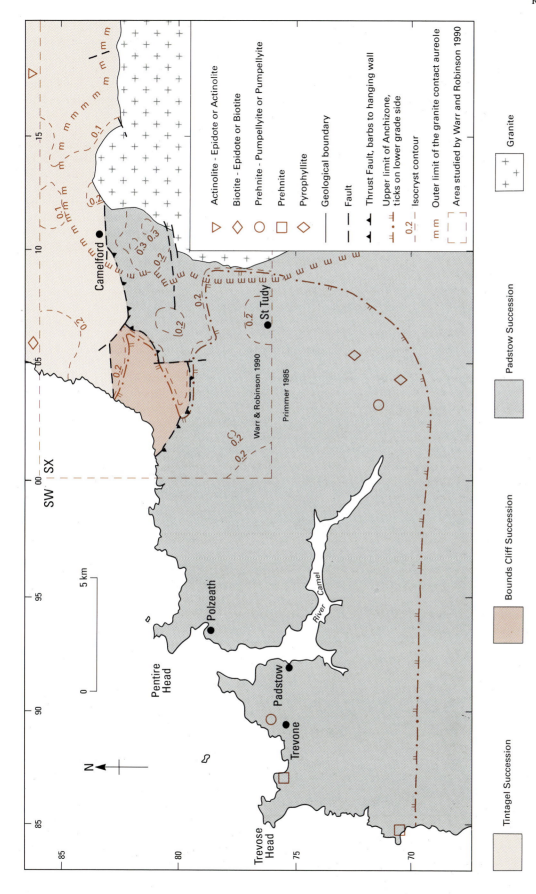

Figure 41 Low-grade metamorphism in the district (after Primmer, 1985a, fig. 1, and Warr and Robinson, 1990, fig. 3).

Liou et al. (1985), and indicates conditions of 300° to 320°C and 1.8 to 2.4 kb pressure. Secondary mineral assemblages in the Helland [SX 074 710] and Cardinham [SX123 687] areas are composed of quartz-albite-actino-lite-prehnite-carbon-white mica-sphene- (epidote-clino-zoisite)-iron (titanium) oxides and sulphides. Albite-quartz-calcite probably developed as interstitial phases from the breakdown of calcic plagioclase in ophitic texture with clinopyroxene (titanaugite); some calcite may also have developed from the breakdown of apatite. Acti-nolitic amphibole pseudomorphs after clinopyroxene developed with c-axis orientation in the S_1 plane.

Illite crystallinity

In terrain where argillaceous rocks predominate, mea-surement of changes in illite crystallinity (IC) provides an indication of variation in grade at low temperatures of metamorphism. The results of an IC survey of the Camelford area, conducted during the present investi-gation (Warr and Robinson, 1990), are summarised in a contoured isocryst map (Figure 41). Two areas of regional metamorphism have been identified in the district: an area of epizone metamorphism to the north with IC values between 0.21 and 0.08 $°\Delta~2\theta$, which includes the Tintagel High Strain Zone, and an upper anchizone area extending to cover most of the central and southern parts of the district. Within the anchizone, the pelites consist predominantly of the 1M mica polytype, chlorite and minor quartz: the pelites of the epizone consist largely of 2M illite, chlorite, minor feldspar and quartz, with some porphyroblasts of chlori-toid and feldspar.

The anchizone/epizone boundary, taken at 0.21 $°\Delta$ 2θ (Kisch, 1980) has been interpolated on Figure 41. It trends east-south-east for 2.5 km from Jacket's Point [SX 033 829] to Westdowns [SX 062 827], veers south-west towards Trewigget [SX 028 800], and resumes an east-south-east trend towards Bodmin Moor. Within 0.5 km of the granite, it is deflected southwards near Michaelstow [SX 089 787], to run parallel to the granite boundary.

The contact metamorphic aureole of the granite appears to overprint the pattern of regional metamor-phism. As a result, the grade in the area of anchizone slates increases towards the granite, indicating that tem-peratures were raised long enough during contact meta-morphism to cause recrystallisation of the illite. This effect is not observed in the area of epizonal slates where recrystallisation of illite had already been effected by regional metamorphism. The broadening of peaks within the aureole of the epizone correlate with areas of calcflintas, and appear to reflect the differing mineralog-ical composition of the pelites.

Anomalies away from the granite aureole do not relate to the distribution of basic igneous rocks. It could be argued that the intrusion of quartz porphyry produced the negative anomaly (high crystallinity) around St Kew [SX 021 768], but there is no surface expression of such an intrusion to explain the comparable negative anomaly at St Tudy.

In the Padstow Unit the regional epizone/anchizone boundary appears to follow strike around the St Minver Synclinorium (Primmer, 1985a; Pamplin, 1990). This pattern of metamorphism, along with the occurrence of pre-S_{1n} chlorite/paragonite stacks described above, was used by Warr et al. (1991) to suggest that the epizone/anchizone boundary was folded during the formation of the regional (D_1) synclinorium, and that the metamor-phism was attained within the Trevone Basin before the onset of folding and thrusting.

Near the southern limit of the district, Primmer (1985a) identified epizone metamorphism, and a sharp epizone/anchizone boundary at the northern end of Watergate Bay [SW 841 650]. Although poorly con-strained inland, the epizone/anchizone boundary repre-sented by Primmer approximates to the northern limit of the Bedruthen Formation, except east of the River Camel where the boundary swings north-eastwards towards St Breward [SX 097766] to cut across the Trevose Slate (Figure 42). It could be significant that structures north-west and south-east of this line show contrasting strike directions (p.67). Pamplin (1990) demonstrated that there is no change in anchizone meta-morphic grade across the Padstow confrontation.

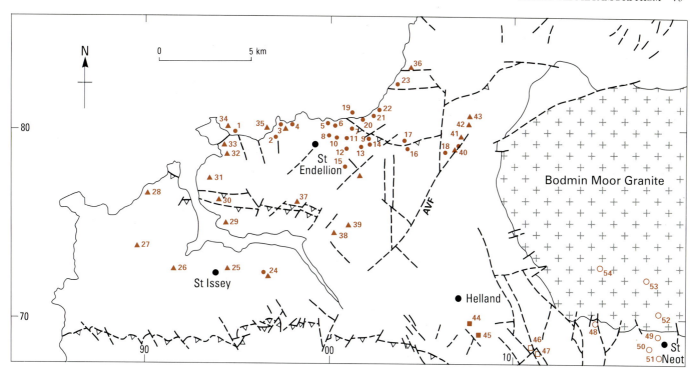

Figure 42 Distribution of mineralisation and mines within the district:

1	Coppinger's Cove	20	Bounds Cliff (south)	39	Wheal Barbara
2	Port Quin Antimony Mine	21	Wheal Sampson	40	Pengenna
3	Port Quin foreshore	22	Ranie Point	41	Treburgett
4	Reedy Cliff/Roscarrock	23	Jacket's Point	42	Whitewell
5	Port Isaac	24	Penhale	43	Treroosal
6	Port Gaverne	25	Legossick	44	Clerkenwater Mines
7	Trewethart	26	Tregonna	45	Penbugle
8	Wheal Thomas	27	St Merryn	46	Deviock Mine
9	Treore Mine	28	Gunver Head	47	Wheal Glynn
10	Trewetha	29	Porthilly	48	Treveddoe
11	Bodannon Mine	30	North Porthilly	49	Wheal Trevenna/Wheal
12	Tresungers Mine	31	Trebetherick		Robins
13	Poltreworgey	32	Polzeath	50	Goonzion
14	Pendoggett	33	Pentireglaze Mine	51	Wheal Mary
15	Trevinnick	34	Pentire Mine	52	Wheal Hammett
16	Tregildren Quarry	35	Gilsons Cove	53	Colliford Downs
17	Tregeare	36	Tregardock Mine	54	Wheal Esther/North Wheal
18	Trelill	37	Treglyn		Providence
19	Tresungers Point	38	Tregorden		

SEVEN

Quaternary

Quaternary deposits occur over much of the district but are especially common on the outcrop of the Bodmin Moor granite. Although there is no evidence in the district to indicate that it was overridden by ice during the Pleistocene glacial phases, extensive head deposits indicate that it was subjected to a periglacial climate at least once during that period. The regional evidence suggests that the southernmost limit of the last Pleistocene (Devensian) ice-sheet was along the north coast of Cornwall. Pleistocene river terrace gravels occur inland, but they show no clear chronology; patches of raised beach deposits, of probable Pleistocene age, occur around the Camel Estuary. Recent deposits in the district are represented by peat on Bodmin Moor, widespread alluvium in the valleys of the major streams, estuarine alluvium in the lower reaches of the River Camel, and an extensive expanse of blown sand which may range in age from Pleistocene to Recent.

HEAD

Deposits mapped as head occur in valley bottoms and on weathered hilltops throughout the district. The deposits vary in thickness up to 2 m, and their composition is dependant on that of the underlying solid rocks. Exposures in head deposits occur along the coast, particularly in the area around the Camel estuary. The Bodmin Moor granite is almost wholly covered by a veneer of head deposits; only the thicker patches are shown on the geological map. Outcrops of granite are restricted to the tors.

On the south side of the Camel estuary, west of Tregunna House, there is a 100 m-long exposure of head [SW 9615 7000] in a cliff section. This head comprises up to 0.4 m of clayey head with small quartz and slate fragments. It overlies a boulder bed, up to 1 m thick, consisting of subangular rock fragments and boulders in a clay matrix, resting on frost-shattered grey slate. Rock fragments within the boulder bed are up to 0.3 m in diameter, and include granite (probably from Bodmin Moor), basalt, dolerite, quartz, quartz-porphyry and grey slate, all of local origin. The granite boulders, with a probable source over 15 km away on Bodmin Moor, suggest a complex origin for the deposit. Reid et al. (1910) suggested that the granite boulders may have been transported by floating ice, and were deposited in a delta laid down where a rapidly flowing river met a tidal estuary. Scourse (1985) correlated the deposit with Arkell's (1943) Trebetherick Boulder Gravel; he also argued that it may have been derived from river ice, and was subsequently redistributed by solifluction. Clarke (1973) suggested that the gravel was of glacial origin, and could be till in a lateral moraine.

A similar sequence has been recorded in other coastal sections in the district. A 100 m-long section near Lellizick [SW 913 773], near the mouth of the Camel estuary, contains two similar beds, although the boulder bed does not contain granite in this locality. Similar sections also occur north of St George's Cove [SW 919 765], and at Gun Point [SW 919 766].

Scourse (1985) suggested dates of less than 29 000 years before present (BP) for the boulder bed at Tregunna, and less than 15 900 BP for the overlying stony clay, on the basis of radiocarbon and thermoluminescence analytical techniques. Both beds therefore fall within the Devensian Stage of the late Pleistocene.

On the Bodmin Moor granite outcrop, the head deposits can be divided into granite regolith, peaty valley fill and granite boulder head. A mixture of regolith and boulder head has been mapped as head on the flanks of steep-sided valleys on the moor, and locally outside the granite outcrop.

The granite regolith is composed of coarse crystal gravel, formed by the mechanical disintegration or fracture-weathering of the granite into its constituent minerals. This process is aided by the freezing and thawing of meteoric waters circulating in the well-developed joint-system near the surface. Deposits over 14 m in thickness have been recorded. Crystal gravel, over 3 m thick, is exposed in the eastbound road cutting of the A30 near [SX 177 706], where the close relationship of the fracture-weathering to the horizontal and vertical joint-systems is clearly displayed. This exposure displays, between the crystal gravel bodies, the characteristic corestones, which closely resemble granite tors in form.

It seems likely that the summit tors of the moor were exhumed from a regolith formed by deep Tertiary chemical weathering, as described by Fookes et al. (1971) for Dartmoor. The lower summit tors and valleyside tors were formed within a later regolith formed by fracture-weathering in Pleistocene times (Te Punga, 1957; Palmer and Neilson, 1962).

Peaty valley fill is an alternation of peat with gravelly hillwash derived from the granite regolith. The gravels have been locally reworked by streams, and carry a variety of channel deposits. Peaty valley fill deposits occur in all the major rock basins and depressions on Bodmin Moor; its thickness is difficult to determine.

Granite boulder head consists of spreads of large granite boulders, set in a fine-grained sandy, commonly peaty, matrix, and is widely distributed over the moor. The spreads are concentrated around the tors, from which they were derived by periglacial freeze/thaw processes. The large extent of the boulder fields suggests that solifluction rafting caused their transport downslope from the shrinking tors.

In the south-west part of the district, Clarke (1971) recorded a section at Porth Mear [SW 847 717] where head on the valley side passes into 'lacustrine' clay with peat layers in the valley centre. The boundary between solid and head is affected by frost action.

RAISED BEACHES

Clarke (1963) and Everard et al. (1964) described strand levels and rock platforms at various levels in the Camel Estuary area, ranging from 4.6 to 46 m (15 to 150 ft) above OD. The latter authors referred deposits between sea level and 7.6 m (25ft) OD to the 'Patella' raised beach, and near St George's Cove [SW 9195 7660] deposits thought to be of this age consist of angular greyish green slate fragments with rounded quartz and dolerite pebbles in a sand matrix. Fragments of *Patella vulgata* from this sand matrix were submitted to Professor H P Sejrup, of the University of Bergen, Norway, for amino analysis. The results indicate an age falling within that of the last interglacial period in Devon.

In the south-western part of the district, remnants of raised beach occur at two localities, on a 10 m platform in the south-eastern corner of Treyarnon Beach [SW 8585 7386], and in the low cliffs backing the southern end of Booby's Bay; in both cases they are largely obscured by blown sand. The Treyarnon deposit is coarse grained with rounded clasts of local origin; the Booby's Bay deposit comprises thin layers of partially cemented fine gravel in coarse-grained sand. Similar deposits were recorded by Clarke (1967) in the cliffs of Mother Ivey's Bay. Semiconsolidated and cemented beach deposits, with a crude bedding dipping gently eastwards, also occur at a lower level in Harlyn Beach [8765 7545] and Little Cove [8645 7597]. These are just below and above high water mark. Clarke referred to the Little Cove deposit as sandrock and as an old beach deposit extending from about 1m below to 3 m above the present strand line.

RIVER TERRACE DEPOSITS

Small areas of sand and gravel have been mapped within the district. A tract, nearly 1 km long and up to 150 m wide, lies in the upper reaches of the River Allen, south of Delabole [SX 069 818 to SX 071 811]. Alluvium formed in stream and river valleys throughout the area, and generally comprises fine sand, silt and mud, commonly with gravel towards the base.

PEAT

Peats are widespread on Bodmin Moor, the thickest sequences overlying peaty valley fill, and have been extensively exploited for fuel. Lenses of hillwash gravel are characteristic of these basin peats. Peat deposits are usually less than 3 m thick, but Pattison (1847) recorded a thickness of 4.3 m (14 ft) in the Fowey valley.

Analysis of pollens from the peats (Conolly et al. 1955; Brown, 1977) has resulted in a detailed picture of the last 13 000 years of environmental change (Caseldine, 1980). The first peats were laid down in the latest Pleistocene (about 13 000 to 10 000). In the early Holocene (about 10 000 to 5000 BP), temperatures rose, and woodland covered much of Bodmin Moor, although the upland areas remained as grassland and heather moorland. Since about 5000 BP woodland cover has declined (Brown, 1977), and the climate has become cooler and wetter.

ALLUVIUM

Extensive alluvial deposits of fine brown silt are developed immediately upstream of the tidal part of the Camel Estuary, in the valleys of the Camel and the Amble. Boreholes show a thickness of 20 m of alluvium on the western side of the Camel, and 3 to 5 m on the eastern side. The sequence is generally gravelly at the base, with sand and silt above. Clarke (1980) suggested that the alluvial plain of the River Amble at Trewornan [SW 984 745] was a lake flat indicating glacial impounding. Scourse (1985), however, showed that the alluvium is Flandrian in age.

BLOWN SAND

Blown Sand deposits extend for about 4 km^2 on the eastern side of the Camel Estuary, forming the St Minver Lowlands. Sections were recorded at Cassock Hill [SW 928 761], Porthilly [SW 939 754], and Porthilly Cove [SW 936 754]. The sands are largely composed of shell fragments, and are commonly carbonate-cemented. Merefield (1989) suggested that the source of the material was the carbonate-rich sands of the Camel estuary, the coastal beaches and reworked raised-beach sands. The blown sand deposits can be divided into a lower, cross-bedded part with slate fragments (up to 2 m thick) and an upper part that consists of sand and silty sand with shell fragments. Arkell (1943) suggested that the lower part was older than the local head deposits and therefore Devensian in age, and that the upper part was Holocene in age.

Many of the inlets, forming the mouths of the valleys which reach the coast south of Trevose Head, are backed by blown sand separating modern beach deposits from head and alluvium. The blown sand acts as a dam producing marshy lacustrine areas in the lower reaches of some of the valleys. The major deposit of blown sand extending from Booby's Bay and Constantine Bay eastwards across the Trevose Head isthmus to Harlyn Bay reaches 30 m above OD. Grassed and cultivated dune forms are recognisable. The sands are quartzose, rich in shelly debris, with complete gastropods and patellids present.

EIGHT

Economic geology

METALLIFEROUS MINERALISATION

The earliest records of the production of metallic ores in the district are from the 16th century, and include evidence of lead and antimony workings (Donald, 1950) in the St Issey and St Endellion areas respectively. Tin and copper production is documented from the St Neot area during the same period (Pascoe, 1945). Reid et al. (1910) described the extensive alluvial tin workings of Bodmin Moor, and it is known that these were in production between the 14th and 15th centuries, although there is no record of yield.

More recent workings in the district, mostly of vein deposits, were of moderate or small size and production was intermittent. The development and output of these small mines reflects the periods of their economic viability which lasted through the 19th century in the St Neot area (Hamilton Jenkin, 1966), and from the second half of the 18th to the end of the 19th centuries in the St Endellion and St Issey areas.

The district is particularly well known for its production of antimony, which was formerly worked from a number of small mines in the area from Wadebridge and Port Isaac towards Bodmin Moor. Special interest in the extraction of antimony occurred during the Napoleonic Wars (Russell, 1949) and persisted sporadically thereafter.

Mining operations for tin and copper produced small tonnages of ore at intervals throughout the 19th century. The largest mine, Treveddoe [SX 1515 6958], produced ore into the early years of the present century (Dines, 1956); it was reopened, although without recorded production, in 1943. No metalliferous mine is now working in the district. As well as vein mineralisation of several types, stratiform mineralisation is present in Devonian and Lower Carboniferous rocks, in the Barras Nose, Trevose Slate and Staddon Grit formations.

The principal metalliferous mineral occurrences are shown in Figure 42; the sequence of mineralisation is summarised in Table 10.

Stratiform mineralisation

Stratiform disseminations of simple sulphide assemblages have been identified in several lithologies in the Staddon Grit, Trevose Slate and Barras Nose formations. All these occurrences are minor and mainly composed of FeS_2 (pyrite + marcasite ± pyrrhotite FeS) with subordinate chalcopyrite, sphalerite, galena and arsenopyrite. These minerals developed early in the sediments, and as replacement deposits in tuffaceous metasediments and metabasites, during diagenetic and regional metamorphic processes. They were mobilised during complex,

Table 10 Classification and chronology of ore deposits of the district.

Metal assemblage	Associated minerals	Type of orebody: those marked (E) have been exploited	Chronology
Fe-As-Cu-Zn-Pb-Sb	Quartz	Minor stratiform bodies	Penecontemporaneous and diagenetic: modified by Variscan deformation
Sb-As-Au	Quartz-carbonates	Shear and fracture vein (E)	Syn-Variscan deformation; pre-granite
Fe-Cu-Zn-As	Fe and Ca silicates	Skarn	Synchronous with granite emplacement
Sn-Cu-W	Quartz-tourmaline-chlorite-fluorite	Vein and replacement (E)	Granite-related
Pb-Zn-Ag	Quartz-carbonates fluorite-barite	Crosscourse vein (E)	Post-granite
Fe	Quartz	Crosscourse vein (E)	Post-granite
Sn	Quartz-tourmaline	Placer (E)	Tertiary and Quaternary weathering and transport

multistage ductile and brittle deformation. Sulphide mineralisation in calc-silicate hornfels and metabasites within the thermal aureole of the Bodmin Moor granite is also present and is attributed to concentration during contact metamorphism. A small manganese deposit was worked at Gutt Bridge [SW 9767 7529] within the Trevose Slate, but details of its exact nature are uncertain.

The highest concentrations of sulphides occur in the black carbonaceous slate of the Barras Nose Slate. This formation also hosts sulphides, particularly at the junctions between volcanic rock and slate. In the fault-bounded sequence exposed at Tregardock Beach [SX 0408 8424], sulphides occur in several texturally distinct modes in the meta-sediments. These include pyrite-marcasite laminae, pyrite-arsenopyrite porphyroblasts, phosphatic- carbonate-sulphide concretions and irregular pyrite segregations which display selective distribution controlled by lithological type and structural site. A paragenetic scheme for the strata-bound mineralisation in the Barras Nose Slate at Tregardock Beach is given in Table 11.

Laminated sulphide occurs at shale/siltstone junctions which acted as permeability boundaries within the sediment. A poorly defined mineral zonation within the laminae consists of an internal zone of pyrite, and an external zone of marcasite. Minor chalcopyrite-sphalerite-galena-arsenopyrite-pyrrhotite segregations also occur in sites peripheral to mineral-rich laminae.

Porphyroblasts of pyrite-arsenopyrite are restricted to primary siltstone/secondary carbonate units which are interbedded with the black slate. They occur as isolated

Table 11 Paragenetic scheme for the stratabound deposits in the Barras Nose Slate at Tregardock Beach.

Mineral	Pre D$_1$	D$_1$	(a) D$_2$ (b)	D$_3$
Marcasite	——	—		
Pyrite	——	—	——	—
Chalcopyrite	—	—		
Sphalerite	—	—		—
Arsenopyrite	—	—	——	
Galena	—		—	——
Quartz	—		——	—
Phosphate minerals	——			
Siderite	——			—
Epidote Chlorite/ white mica			——	— —

euhedra containing inclusion trails of detrital and organic material which define the S$_1$- spaced fabric, indicating the overgrowth of the early cleavage by sulphides and the sealing of the primary porosity.

Sulphide segregations in black carbonaceous slate and volcaniclastic rocks of the Barras Nose Slate and Trevose Slate at Tregardock Beach and Reedy Cliff [SW 9790 8107] are associated with minor quartz and carbonate gangue minerals. These elongate lenses attain 15 cm in length, and display a co-planar relationship with the bedding of the host rock.

Lower in the succession, scattered occurrences of minor disseminated pyrite are developed in thinly bedded siltstone/sandstone/mudrock units of the Staddon Grit on the northern foreshore of Mawgan Porth [SW 8450 6800]. The sulphides are confined to the coarser-grained clastic rocks and are virtually absent from the mudrocks.

Calcareous pelites in the thermal aureole of the Bodmin Moor Granite acted as hosts to disseminated stratiform concentrations of sulphide during metamorphism. Reconstitution of the rocks permitted textural and mineralogical changes, involving the development of compositional banding, and growth of a calc-silicate mineral assemblage comprising diopsidic pyroxene, actinolite (tremolite), epidote, idocrase, garnet, quartz and albite. Sulphide mineralisation (pyrrhotite-pyrite-marcasite-chalcopyrite-sphalerite) is stratiform in character and co-planar with compositional banding. Richly mineralised layers form ochreous weathering residues, for example at Helsbury Quarry [SX 0877 7908]. Occurrences are restricted to calcareous pelites within the inner aureole in the St Neot district and adjacent to the western flank of the granite.

At Port Quin [SW 9703 8058] and Port Gaverne [SX 0015 8095] sparse disseminations of early sulphides are enclosed in texturally zoned pyrite-marcasite. These blocky textured secondary sulphides replace thin siltstone and tuffaceous beds, and are associated with secondary quartz and calcite. The sulphide zones are tightly folded with bedding, and quartz fibres have formed during D$_1$ in the strain shadows of pyrite nodules. West of Com Head [SW 9388 8050], a replacement deposit of massive pyrite with subordinate chalcopyrite and sphalerite forms a stratiform layer 0.1 m in thickness in interbedded dark grey slate, basaltic lava and tuff. The mineralisation is co-planar with the bedding fabric and is continuous for about 10 m. Similar occurrences exist in sequences of dark grey slates and basaltic tuffs for example at St Saviours Point [SW 9225 7600]).

Sulphide impregnation of basaltic tuff is widespread within the Barras Nose Slate at Tregardock Beach [SX 0408 8422]. Mixed sulphide assemblages occur, dominated by pyrite-marcasite intergrowths. The occurrence of disseminated tetrahedrite in an intensively altered and silicifed tuff [SX 0418 8440] indicates that stratiform mineralisation hosted in volcanic rocks could be the source of the antimony, later redistributed in cross cutting quartz-carbonate-sulphide veins.

Irregular sulphide segregations commonly reside in lenses of basaltic lava, which display tectonic boundaries with the enclosing sedimentary rocks. Sediment–basalt junctions were also sites of sulphide deposition during deformation, suggesting that fluid-assisted redistribution of sulphides during deformation was important. A similar host relationship in a black slate/lava sequence in the Trevose Slate at Port Quin [SW 9685 8084] suggests that, in addition to the common disulphides, lead-antimony sulphosalts were redistributed during deformation.

Hornfelsed metabasites in the contact aureole surrounding the Bodmin Moor granite, for example at Corner Quoit [SX 1226 7078], locally contain sulphide disseminations. Common constituents include pyrrhotite, pyrite, chalcopyrite, sphalerite and arsenopyrite, which occur interstitially in a plagioclase-quartz-epidote-hornblende-biotite metamorphic assemblage. The predominance of the pyrrhotite over pyrite suggests a reduction of earlier formed sulphides during metamorphism. Pyrrhotite is subordinate to pyrite-marcasite beyond the thermal aureole.

Vein mineralisation

ANTIMONY-ARSENIC-GOLD MINERALISATION

Antimony-arsenic-gold mineralisation occurs as vein systems within the volcanic sequences of the Trevose Slate, Harbour Cove Slate and Jacket's Point formations in the northern part of the Trevone Basin. The St Endellion area was the principal location of antimony production in south-west England (Dewey 1920), for example Wheal Boys (Trewetha) [SX 0050 8010] produced 95 tons of antimony ore between 1774 and 1776 (Pryce, 1778). During the period of greatest activity the mine was stated to have returned 100 tons of ore, with 40 per cent antimony, per annum (Hogg, 1825). Subordinate quantities were also produced from Legossick [SW 9477 7240] and Penhale [SW 9615 7230] mines in the St Issey district. An attempt to reopen Trevinnick and Treore mines in the early years of the present century did not result in any recorded production.

The mineralisation appears to be deformation-related; quartz gangue crystals show strain. Fluid-inclusion geothermometry on quartz crystals (Clayton et al., 1990) suggests that the ores formed at temperatures of between 280° and 315°C from low salinity NaCl-dominated brines.

Gold occurs in both the hanging-wall and foot-wall of a thrust (D$_{2b}$ age) exposed in the coastal section at Tresungers Point [SX 0088 8119]. It has also been recorded from former mines at Treore and Trevinnick, and from veins exposed at Bounds Cliff [SX 0230 8144] and Port Quin [SW 9709 8058]. Although the significance of this thrust to the distribution of mineralisation remains uncertain, fluids associated with it acted as agents in the remobilisation and redistribution of sulphides. Early mining records indicate that deposits were small and impersistent, both laterally and vertically (Dewey, 1920).

A generalised paragenetic scheme, Table 12, indicates an early predominance of arsenopyrite-pyrite succeeded by the growth of lead-antimony sulphosalt assemblages. The Stage 1 arsenopyrite-pyrite is commonly brecciated and cemented by Stage 2 sulphides; narrow fractures in the earlier assemblage may be filled with gold. Quartz with rare tourmaline (dravite) and sphene, are the predominant gangue minerals of Stage 2 and quartz displays highly strained microfabrics. Stage 3 of the paragenesis involved complex deposition of sulphosalts as intergrowths, consisting of tetrahedrite, bournonite, jamesonite, boulangerite and plagionite with associated stibnite. The lead-antimony and copper-antimony sulphosalts fill inter- and intragranular fractures in quartz, and may be intimately associated with the deposition of carbonate (dolomitic siderite). Oxidation of these sulphosalts to covellite, stibiconite, and bindheimite is associated with gold deposition, developed in altered zones of sulphosalt minerals e.g. covellite rims

Table 12 Generalised paragenetic scheme for antimony–arsenic–gold mineralisation in the district.

	Stage 1	Stage 2	Stage 3	Secondary
Quartz	——————	—		—
Arsenopyrite	— —			—
Pyrite	——			
Chalcopyrite		—		
Tetrahedrite			— —	Covellite
Sphalerite		—		
Galena			—	
Bournonite			—	
Jamesonite			—	Bindheimite
Boulangerite			—	
Plagionite			—	
Stibnite			—	
Gold			—	Stibiconite
Siderite/calcite			——	
Chlorite		—		
White mica		—		
Quartz deformation events				
Sub-grain formation	——			
Stylolitisation	—			
Linear fracturing	—			
Vugh development			—	—

on tetrahedrite. The silver concentration of tetrahedrite was found to range up to 21 weight per cent Ag. The content of galena within these veins is subordinate, a proportion occurring as a possible breakdown or reaction product between members of the sulphosalt assemblage.

Tin-tungsten-copper mineralisation

The principal hydrothermal vein deposits in the Bodmin Moor Granite are sparsely distributed, mostly in the southern part of the pluton. Typical lode systems are oriented 080° to 090°. Those which have been worked are mostly situated close to the granite/country rock contact and have yielded small tonnages of tin and copper (Dines, 1956). Some of the veins bear small quantities of wolframite, but there is no record to show that tungsten was produced. The central part of the pluton appears to be almost barren, although extensive former alluvial workings for tin suggest that vein systems may have been emplaced at a higher level in the pluton and subsequently eroded. Little is known of the grades and yields of these ancient alluvial workings. An account (Austin et al., 1989) of archaeological work in the St Neot River valley deals with medieval mining in the area now covered by the waters of the Colliford Reservoir, and includes a section (Scrivener, 1989) on the geological features around the former West Colliford Mill [SX 177 713]. There, cassiterite-bearing alluvial gravels were worked and a low-grade vein was mined; the latter includes an irregular trench-like excavation [SX 1751 7120] extending east-north-east–west-south-west over 350 m, with pits indicating limited underground development. The granite wall-rocks of the trench are heavily kaolinised and sericitised with local reddening and tourmalinisation. Close to the vein, the granite is altered to a greisen of muscovite, quartz and minor topaz. Specimens of vein material collected from the open working and from spoil dumps in the vicinity of the mill carry mostly low tin values (less than 0.1% Sn) with rare specimens of higher grade (up to 1.5% Sn). The ore consists of very finely crystalline tourmaline enclosing brecciated fragments of quartz-muscovite greisen, local patches of haematite and other iron-oxide minerals. Cassiterite is present as a minor phase and occurs in broken, irregular crystals showing, in thin section, concentric colour zoning from pale yellow-brown to dark red-brown. The cassiterite appears to be associated with the muscovite-quartz greisen rather than with the tourmaline vein-filling, and to have been included in the latter as brecciated fragments. The sequence of mineralising events appears to have been:

Growth of brown tourmaline aggregates in the granite host
Fracturing and formation of greisen bodies, possibly bordering quartz-cassiterite veins
Fracturing and brecciation of the greisen bodies; vein filling with finely crystalline grey-blue tourmaline
Interstitial cavities filled with quartz and haematite
Fracturing and development of quartz-filled cross veins (crosscourses)

A similar pattern of mineralisation is shown at an open working [SX 1564 6968] 400 m east of Treveddoe and known as Good-a-Fortune. The walls of the open working and specimens collected from spoil heaps nearby show greisened granite cut by tourmaline veins of variable thickness, which commonly enclose brecciated fragments of quartz and greisen. Thin sections show the greisen to be composed of aggregates of irregular, interlocking quartz grains with undulose extinction, together with ragged, strained crystals of muscovite up to 1.5 mm across, and corroded irregular crystals of brown tourmaline. Aggregates of fine muscovite (sericite) are present throughout; they enclose and are enclosed by quartz. Scattered cracks are filled by iron oxide minerals. The veins are composed of brecciated fragments of quartz, greisen and minor brown tourmaline enclosed by finely crystalline, acicular pale brown to grey-blue tourmaline prisms forming a felted mass. Rare small crystals of red-brown cassiterite are scattered at the margins of some of the fine tourmaline aggregates.

Lode systems in the country rocks of the district that carry tin-tungsten-copper mineralisation were developed in a zone close to the southern margin of the Bodmin Moor granite. Within the dark grey cordierite pelites and calcsilicate hornfels of the metamorphic aureole, the strike of the lodes varies from east–west to east-north-east-trending, parallel to a suite of quartz porphyry (elvan) and aplite dykes. Mineral assemblages in the veins comprise cassiterite, wolframite, pyrite, arsenopyrite, quartz and tourmaline, with later chalcopyrite and sphalerite. Haematisation is common and represents the latest effects of circulating mineralising solutions.

The principal production in the district was centred in the St Neot area at Treveddoe [SX 1515 6958], Trevenna Wood [SX 1830 6867] and Wheal Mary [SX 1886 6734]. The last included Wheal Sisters and Ambrose Lake Mine and was of greatest importance, producing 13 889 tons of 6 per cent copper ore between 1828 and 1864 and 144 tons of black tin and 150 tons of 6.25 per cent copper ore between 1872 and 1876. It also returned 80 tons of 80 per cent lead ore and 320 ozs of silver from a crosscourse encountered at depth (Dines, 1956). Workings of lesser importance operated during the mid-nineteenth century, along the western flank of the granite around St Breward and Blisland, confined to the ground east of the River Camel. Extensive old surface workings across Goonzion Downs [SX 1770 6770] exploited tin and copper from veins and from mineralised elvan dykes. The former production from these workings is unknown; recent (1979) exploration was not encouraging.

Lead-zinc-silver (crosscourse) mineralisation

Lead-zinc-silver workings in the district were concentrated on north–south-oriented vein systems 'crosscourses' with the exception of Treburgett Mine [SX 0573 7988] which worked a north-east–south-west system parallel to the Allen Valley Fault. These deposits persist intermittently over greater distances than the antimony-arsenic-gold mineralisation. The largest mines existed at Treburgett, which produced 2180 tons of 75 per cent lead ore, 44 tons zinc and 9530 ozs silver between 1871–1881 (Dines

1956), and along the Pentire/Pentireglaze vein system. Both these mines and many smaller deposits, were located within the area of outcrop of the volcanic rocks of the northern zone of the Trevone Basin. The predominant host lithologies are black and dark grey carbonaceous slates and metabasites, and include the Lower Carboniferous Tintagel Volcanic Formation. In contrast with the fabrics in antimony-arsenic-gold veins, the quartz gangue crystals of the crosscourses show unstrained microfabrics, and elsewhere lower temperatures have been obtained from fluid-inclusion thermometry of crosscourse mineralisation (Shepherd and Scrivener, 1987). The association of gold with lead deposits as at Port Quin, [SW 9709 8058] is uncommon. The working of lead and silver in the Helland and Cardinham areas occurred from the mid-nineteenth to early twentieth centuries (Hamilton Jenkin, 1966, 1970; Dines, 1956). The deposits are located in lode systems with north–south orientation and are less abundant than the lead-zinc-silver deposits in the northern zone of the Trevone Basin. Their economic importance was subordinate to that of the tin-tungsten-copper ventures in this sector.

Two principal deposits, containing galena, chalcopyrite, pyrite, sphalerite, chlorite and carbonate are located at Clerkenwater [SX 0690 68880], where the main gangue minerals are quartz and fluorite, and in Hurtstocks Wood [SX 1113 6750], with a gangue of quartz and barite. Wheal Glyn [SX 115 6775] was also reported to have produced 80 tons of lead ore (Hamilton Jenkin, 1966). Much of the galena was described as argentiferous (Hamilton Jenkin, 1970; Dines, 1956) suggesting the occurrence of mineral inclusions of silver bearing sulphosalts.

CHINA CLAY

Kaolinisation of the Bodmin Moor Granite has been sufficiently intense for it to have been worked at many localities in the district. Currently china clay is worked at Stannon [SX 128 812] and Park [SX 194 708]. All the china-clay workings in the district, past and present, lie in the western part of the Bodmin Moor granite. Most of the china clay produced in south-west England comes from the St Austell and Dartmoor granites. The Bodmin Moor granite is the third most important producing area, with an average production in recent years of around 200 000 tonnes per year, all from English China Clay International Ltd (ECC). Most of the current production is used as a paper filler. Total annual value of the production is currently (1990) just over £10 million, making it the most important mineral product in the district.

There are abandoned china-clay workings at Durfold [SX 119 738], Carwen [SX 110 737], Temple [SX 137 731], Glynn Valley [SX 144 718], Cardinham Moor [SX 128 722], Gazeland [SX 166 698] and Hawkstor [SX 150 747]. All of these workings were abandoned prior to the First World War, with the exception of Hawkstor and Glynn Valley. Hawkstor was closed in the early 1970s, largely because of high operational costs and the low quality of the remaining reserves. Glynn Valley was

closed between the wars, but a significant quantity of kaolinised granite was found at this location by drilling in the early 1970s.

Early trials in the Stannon area were made in the 19th century: Poldue pit to the north [SX 131 816], which shipped china clay via Boscastle in the 1850s, and Hennward Pit [SX 127 808] which was separate and to the south of the present Stannon Pit and operated until the 1930s.

Trials for china clay have taken place in many areas and it is known that many of the larger marshy hollows ('slads') are underlain by kaolinised granite, albeit of poor quality. Drilling by ECC in the marshy areas west of Buttern Hill [SX 167 821 and SX 177 802] proved this 'moor' type of shallow, poor-quality kaolinised granite, and an old trial at Deweymeads [SX 168 732], now submerged below the waters of Colliford Lake, showed pink- and red-stained kaolinised granite. Trials for china clay were also reported at Merrifield [SX 145 729] and Colquite [SX 163 738], (Barton, 1966). Drilling for a proposed dam at Lamelgate in the Fowey valley, in the adjacent Tavistock district proved a similar type of poor- quality china clay.

In the late 19th and early 20th centuries, transport of coal to the china clay-works and of china clay away from the works on Bodmin Moor presented a considerable problem. Salt-glazed pipelines were laid to connect the china clay-pits to the drying units alongside the railways of the region. The pipelines from Stannon to Wenford and from Park to Moorswater (near Liskeard) are still in use.

The character of the kaolinisation on Bodmin Moor has been deduced from a combination of drilling, resistivity surveys, photogeological interpretation and geological inspection of the faces in the active pits. Two types of kaolinisation have been recognised:

'Moor' kaolinisation, usually yielding china clay of little commercial value, occurs as widespread, thin, kaolinised zones underlying most of the marshy hollows on the moor. 'Deep' kaolinisation occurs mostly at the intersections of major lines of fracturing of the granite. In the working pits, the fracture zones are marked by non-tourmaline-bearing subvertical quartz veins, commonly with iron staining and with intense kaolinisation of the granite on either side. At Park Pit [SX 194 708] an east-north-east-trending set of fractures predates, and is displaced by, a north–south set. The north–south set appears to correspond with the Stage IV cross-course alteration stage of Bristow (1990). The east-north-east-trending set may represent an early phase of Bristow's Stage IV, from a time when the minimum horizontal stress direction was north–south (as in Stage III).

The clay produced by 'deep' kaolinisation at these major lineament intersections is thicker and of better quality than the 'moor' kaolinisation, and deposits such as those at Park and Stannon are of similar quality to the china-clay deposits in the St Austell area.

Kaolinisation of the Bodmin Moor biotite-bearing granite to form white china clay has involved the removal of much iron. Iron lodes, which may represent some of this displaced iron, occur close to the granite at Devil's Jump [SX 103 800] and south of Helstone [SX 090 807].

At Stannon Pit, a belt of strong discoloration, caused by iron-staining on the south side of the kaolinised zone, may represent iron displaced from the main kaolinised area. In other south-west England granites, such as St Austell, tourmaline acted as a scavenger for the iron and fixed it in a stable mineral which remained unaffected by the kaolinisation (Bristow, 1990), but tourmaline is less common in the Bodmin Moor china clay-pits than in the St Austell pits.

Fehn (1985), Sams and Thomas-Betts (1988) and Jackson et al. (1989) have suggested that kaolinisation in south-west England is associated with the downward limbs of convection cells of radiogenically heated meteoric water that circulated in the granite, with a particular concentration of downward flow close to the granite margin. The positions of the Park and Stannon zones of deep kaolinisation fit this model. The 'moor' kaolinisation could be a later feature, in which weathering was an important element in its formation.

Park Pit shows a number of interesting features. Unlike most of the china-clay pits in the St Austell area, it has only a limited amount of quartz-tourmaline veining and associated greisenisation and tourmalinisation. The shallower areas of kaolinisation to the north and south of the central pit area show typical 'moor' kaolinisation to depths of 20 to 30 m, whereas the central area of the pit shows higher quality kaolinisation to greater depths. Thick, good quality kaolinisation is known from drilling in the area of a natural hollow, east of Park Pit [SX 202 710]. It too seems to correspond with the intersection of north–south and east–west lineaments. South of Park Pit, Northwood Pit [SX 197 702], worked early this century, shows white and variably coloured kaolinised granite.

Stannon Pit, in the northern part of the Bodmin Moor granite, lies on the intersection of three lineaments. An east-north-east-trending lineament extends through the pit area and is recognisable as two parallel quartz veins in the pit, possibly representing a normal fault that downthrows to the north. South of the pit, a north-north-east-trending lineament runs from Stannon Farm [SX 128 804] to Delford Bridge [SX 114 759], and a north-east- trending lineament runs across Harpurs Downs [SX 117 797] to the valley of the stream near Tuckingmill [SX 090 779]. The east-north-east-trending lineament appears to change direction at the intersection with the other two lineaments, trending east–west to the east of Stannon Pit. A dextral movement has been suggested for this fault, which suggests that the intersection could be a position of structural weakness, that was exploited by the downward-flowing water. The deepest kaolinisation lies at this intersection; drilling has shown that there is a substantial area which has over 60 m depth of kaolinisation, much of which is comparable in quality to the better St Austell china clays. There is evidence that clay quality improves with depth, possibly due to change to a less iron-rich parent granite.

QUARRYING

Small abandoned quarries, formerly used for local supplies of walling stone and road metal, occur throughout the district. The walling stone was mostly derived from the slates, the road metal from igneous rocks.

There are several slate quarries in the district, although some are not in continuous production. The Tredinnick Rustic Slate Quarry [SW 935 688] produces slate mainly for walling, fireplaces and other decorative use. The slate is produced from the Bedruthan Formation, and has a density of 1493 kg/m^3. The nearby Cannalidgy Farm Quarry [SW 939 699] produced similar material from the same formation, as does the St Jidgey Farm Quarry [SX 941 698].

The Penquean [SW 955 738] and Camel [SW 953 738] quarries were worked until the early years of the 20th century, providing slate from the Trevose Slate. Bedruthan Formation slates are worked at the Callywith Quarry [SX 086 681], and are sawn for walling, fireplaces and ornamental building work.

Delabole Quarry [SX 075 839] continues to be worked; the slate is hard, durable, of low density, impermeable and has a permanence of colour. It is used largely for roofing and ornamental purposes; crushed slate is used in artificial roofing, plastic, putty and paint, and as an inert filler for fertilisers and, in the past, in cosmetics. Jacket's Point Slate is quarried on a small scale at Lower Tynes [SX 044 819], where roughly cut slate with iron- and manganese-stained surfaces is sold for ornamental purposes.

Multicoloured ornamental slates are also produced from the Tredorn Slate Formation at Trebarwith [SX 068 851], Trecarne [SX 056 854] (intermittent production) and Trevillett [SX 082 881] quarries.

Granite and elvan were formerly extensively quarried on Bodmin Moor, but changes in construction techniques have led to a sharp drop in the demand for dressed stone in recent years. Numerous small quarries were opened on the moor to supply local farm buildings. Most have been infilled but temporary pits to supply rough walling stone are opened from time to time. Tor Down Quarry [SX 093 767], near the western margin of the Bodmin Moor granite, ceased production in the late 1980s.

Two quarries are currently (1991) worked for building stone, both near the western margin of the granite in the De Lank valley, at Hantergantick [SX 1034 7569] and De Lank [SX 1010 7550] (Plate 5b). The granite is hard and durable, and is used for constructions demanding particularly high-strength specifications. The Eddystone, Bishop's Rock, Smalls Rock and Beachy Head lighthouses were all built from granite from De Lank Quarry in the 19th Century. The uniform and relatively fine-grained texture of the rock also make it a valuable ornamental stone. Much of the production from De Lank Quarry is cut and polished on the site. The crushing strength of the De Lank granite is 1280 kgf/cm^2, and the density 2642 kg/m^3; the figures for the Hantergantik granite are 1680 kgf/cm^2 and 2640 kg/m^3 respectively.

Most of the elvans on the moor have been worked in the past for building stone. Their closely spaced jointing makes the rock easier to work than granite: it was mostly used locally for building and walling stone. The complex of elvans in the De Lank valley has been exploited for road aggregate.

SOILS

Five major soil groups and 16 associations have been recognised in the district (Findlay et al., 1984). The distribution of these soils is shown on Soil Survey 1:250 000 Sheet 5 (1984); part of the district is covered by 1:25 000 Sheet SX 18 (1976). The latter covers a representative area from the granite uplands of Bodmin Moor to the country rocks lying beyond the metamorphic aureole (Staines, 1976).

The distribution of soil types in the district (Figure 43, Table 13) reflects the interplay between the bedrock, rainfall and relief. Topographically the area is dominated by Bodmin Moor, where heavy rainfall causes intense soil leaching and the development of podzolic and surface-water gley soils. The topographically lower, slaty country rocks generally yield loamy, stony, permeable soils in which leaching is the main soil-forming process. Groundwater gleying also operates in topographical depressions. The higher land formed by the Lower Devonian sandstones suffers some podzolisation, but surface-water gleying is uncommon.

Lithomorphic soils

Shallow, noncalcareous loamy soils, well drained and with a distinct brownish coloured topsoil are developed in the dissected lowlands bounding the Camel Estuary; one of the driest areas in Cornwall with a mean annual rainfall of less than 1000 mm. The droughtiness of the soils favours arable farming, and winter-sown cereal crops are extensively grown.

Deep calcareous sandy soils are formed on the stable dunes and areas of blown sand backing the dunes at Constantine Bay and on the east side of the Camel Estuary, north of Rock. These soils are droughty, unstable and rarely suitable for agriculture.

Brown soils

Fine loamy, brown, stony, well-drained soils that extend to moderate depths characterise the higher land rising eastwards towards Bodmin Moor and southwards to the Lower Devonian uplands of St Breock Down. They provide excellent permanent pastureland.

Well-drained, loamy, brown earths are developed over the higher land formed by dolerites, basic tuffs and lavas.

Stony, well-drained loams are developed on the Tintagel Volcanic Formation, and principally support permanent grassland.

Well-drained, coarse, loamy, brown podzolic soils, characterise the granitic head developed in the marginal areas of Bodmin Moor, and extend into the higher ground along the river valleys. In normal seasons,

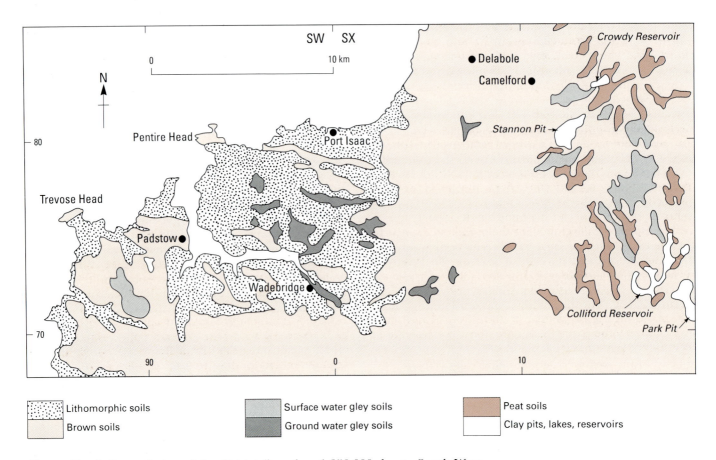

Lithomorphic soils
Brown soils
Surface water gley soils
Ground water gley soils
Peat soils
Clay pits, lakes, reservoirs

Figure 43   Soil association of the district (based on 1:250 000 sheets, South West England: Soil Survey of England and Wales).

Table 13 Relationship of principal soil types to solid formations and drift deposits.

Drift and solid rocks	lithomorphic	brown	surface-water gley	groundwater gley	peat
Blown Sand	+				
Alluvium	+	+		+	+
Peaty Valley Fill		+			+
Boulder Head		+			+
Undifferentiated Head	+			+	
Barras Nose Formation		+			
Bedruthen Formation	+	+			
Harbour Cove Slate Formation	+	+		+	
Jacket's Point Slate Formation	+	+			
Meadfoot Group (undifferentiated)	+				
Polzeath Slate Formation	+	+		+	
Rosenun Slate Formation		+			
Staddon Grit Formation	+	+			
Tintagel Volcanic Formation		+			
Trambley Cove Formation		+			
Tredorn Slate Formation		+			+
Trevose Slate Formation	+	+	+	+	+
Granite		+	+		+
Microgranite		+	+		+

droughtiness presents no problems and the soil is able to support permanent grassland farming.

Free-draining, fine loamy soils are mainly developed in areas above 200 m, typically on the margins of Bodmin Moor and on the lower land along the steeply wooded valley slopes of the rivers Camel and Allen and their tributaries. Such soils have also developed on the Staddon Grit Formation, where repeated cultivation has disturbed the iron pan and reduced the organic matter in the topsoil. Because of the high rainfall, these soils, though well drained, remain moist throughout the year and support permanent grassland, though much is only suitable for rough grazing.

Humic brown podzolic soils, extensively developed on the upland granitic head of most plutons in south-west England, are only present in the district on the highest part of the moor around Brown Willy, and on the steep slopes rising from the Camel valley. They are a coarse, loamy, well-drained soils with dark brown, humus rich topsoil which, on gentle slopes, supports grassland of low grazing value.

Coarse, loamy acidic soils are developed most extensively on the granitic uplands in the district. Their peaty topsoils retain much water, and iron pan within the soil profile creates waterlogging problems. Heavy grazing has suppressed much natural heath vegetation on these soils, and they are now characterised by purple moor grass (*Molinia*) and acid bent-fescue grassland.

Loamy upland soils, with a humus rich or peaty topsoil and iron pan, are developed on the high land formed by the Staddon Grit and Bedruthen formations where the mean annual rainfall is 1000 to 1200 mm. In this area, reclamation by subsoiling has developed grassland used for dairy and stock cattle. Similar soils extend from the north-western margin of the Bodmin Moor granite, across the high-level plateau of Davidstow Moor. Land improvement there has permitted the development of forestry and permanent grassland.

Surface-water gley soils

These seasonally waterlogged soils show a distinct topsoil, but no clay-enriched subsoil. They have been mapped only on the ridge top south-east of Trevose Head, extending from Shop [SW 878 737] towards Rumford [SW 897 710]. Rainfall is currently low is this area, and the gleying may therefore relate to a former wetter climate.

Very acid, granitic soils, with a thin peaty topsoil overlying grey, gritty, sandy loam, have formed on crystal-gravel head. It is the characteristic soil on the broad slopes forming the highest parts of Bodmin Moor. Mostly the soils are permanently wet; intensive stocking has frequently wasted the peaty topsoil and the natural heather growth has been extensively replaced by *Molinia* and bristle-agrostis grassland.

Groundwater gley soils

Fine silty, typical alluvial, gley soils are developed on the alluvium in the lower parts of the Camel and Allen valleys. These soils provide good permanent grassland which is mainly devoted to stock and dairy cattle.

Fine, loamy, typical cambic gley soils occupy seasonally waterlogged ground, in broad depressions and valley bottoms in areas of other well-drained soils. Cambic soils, produced by alteration or removal of mineral matter, have a fine texture and are characterised by red and grey colour mottling. These soils are devoted to permanent grassland.

Humus-rich gley soils, with peaty topsoil over a mottled, coarse, sandy loam, are developed on the stony granitic drift outwashed from Bodmin Moor by the De Lank River. This undrained area is waterlogged for much of the year, and is used for summer grazing.

Peat soils

Raw acid peat soils are well developed in the valley bogs of Bodmin Moor, where they are kept perennially wet by high groundwater. The bogs are only of agricultural importance as rough summer grazing. Peats developed around Crowdy Reservoir [SX 145 835] extend beyond the granite's margin onto the metasediments of Davidstow Moor.

WATER SUPPLY AND HYDROGEOLOGY

There is no major aquifer within the district, and the majority of the district's supply comes from surface water sources. The average annual rainfall varies from less than 800 mm at Stepper Point to over 1800 mm on the high ground of Bodmin Moor; the annual evapotranspiration is around 565 mm. There are three licences for abstracting a total of 10 794 megalitres per annum (Ml/a) from surface sources in the area for public water supply; these apply to Colliford Lake [SX 179 711], the De Lank River [SX 135 765] and Crowdy Reservoir [SX 139 833].

There are no groundwater sources licensed for public supply. However, small supplies are obtainable from most formations (Table 14). Most of the water (298 Ml/a) is abstracted from a total of 167 sources for agricultural purposes, including one abstraction for fish farming. Small quantities are also abstracted for private and domestic water supplies (18 Ml/a) and for china clay processing (4 Ml/a). The district lies within hydrometric areas 47, 48 and 49 and its water resources are administered by the South West Region of the Environment Agency.

There is little difference in the water-bearing characteristics of any of the Devonian and Carboniferous formations; the quantities abstracted reflect differences in the sizes of their outcrop areas rather than their rock-types. Analysis of the yields of 65 boreholes in the district show that they follow a log-normal distribution, with a mean value of 0.6 litres per second (l/s), and a 20 per cent probability of the yield being less than 0.3 l/s. Boreholes are generally between 20 and 40 m deep. Rest water levels are mostly within 5 m of the ground surface. Examples of higher than average yields are a 38 m deep borehole at Menkee [SX 057 726] which yielded 2.3 l/s from Middle Devonian grey slates, and a 152 mm diameter 30 m deep borehole [SX 1586 8633] north-east of Camelford, which yielded 2.0 l/s from the Tintagel Volcanic Formation. In the mid-nineteenth century, the Butter Well [SX 049 680], a spring close to the River Alen near Dunmere, was pumped at 8.8 l/s to a reservoir on the Beacon, from which it supplied Bodmin.

The granite, similarly, provides small supplies. The mean yield of the 10 boreholes in the area is 0.4 l/s, with a 20 per cent probability of a yield of less than 0.2 l/s being obtained. The boreholes are generally 20 to 30 m deep. Higher yields were obtained from a 24 m deep borehole at South Penquite [SX 102 753] which provided 2.3 l/s. Several shallow wells (average depth 4.6 m) near St Breward [SX 1055 7874] provided a maximum yield of 18.4 l/s when constructed in 1893. Rest water levels are generally within 10 m of the surface. The top 3 m or so is often weathered, but below this depth the rock is coherent and does not require supporting.

As water in both the sedimentary rocks and the granite occurs in joints and fissures, yields are dependent on the

Table 14 Quantity of groundwater licensed to be abstracted from various formations in April 1991.

	megalitres per annum
Superficial deposits	14
Lower Carboniferous	5
Upper Devonian	64
Middle Devonian	139
Lower Devonian	25
Tuff	1
Diabase and Proterobase	7
Granite	64
Total	320

Data supplied by South West Region, Environment Agency.

size of the fissure system a borehole intersects, and are often only sustainable for short periods of time, particularly during periods of limited rainfall. As groundwater sources tend to be shallow they are vulnerable to surface pollution. Nitrate concentrations, locally elevated by agricultural practice, may cause problems for domestic supply. The groundwaters have short residence times in the aquifers. Waters from both the shales and the granite are similar in composition to surface waters with low total dissolved solids contents; they may be acidic (pH less than 5.5) and contain suspended matter. Waters from the granite generally have lower dissolved solids contents than those from the country rock. The shallow wells near St Breward had only 69 milligrams per litre (mg/l), while levels of over 210 mg/l were recorded from a spring at Costislost [SX 0278 7015] in the Middle Devonian grey shales of which the majority of the dissolved solids were attributable to bicarbonate. Iron may be present in some groundwaters derived from the shales and sodium and chloride concentrations increase towards the coast.

The radon contents of surface and groundwaters in the granite areas of south-west England have been the subject of many studies. As yet there are no values available for radon concentrations in this district. However, Burgess et al. (1982) have recorded concentrations between 100 and 740 Bequerels per litre in near surface groundwaters and minewaters from the Carnmenellis Granite. Durrance and Heath (1985) concluded that high concentrations of radon in surface streams on the Dartmoor Granite arose from discharge of groundwater on the rising limbs of convection cells.

WASTE DISPOSAL

There are two operational landfill waste disposal sites taking domestic refuse in the district, at Trewarder [SX 015 724] and Bowithick Quarry [SX 072 860]. Several others take construction and demolition waste; sites at Blue Haven, St Clether [SX 182 853] and The Foundry, Wadebridge [SW 984 713] take water- treatment sludge and foundry sand respectively. All the sites are located on the slates, except for that at Blue Haven (Tintagel Volcanic Formation), and are unlikely to pollute groundwater sources.

REFERENCES

Most of the references listed below are held in the Library of the British Geological Survey at Keyworth, Nottingham. Copies of the references can be purchased from the Library subject to the current copyright legislation.

AGRELL, S O. 1939. Adinoles of Dinas Head, Cornwall. *Mineralogical Magazine*, Vol. 25, 305–337.

— 1941. Dravite–bearing rocks from Dinas Head, Cornwall. *Mineralogical Magazine*, Vol. 26, 81–93.

AL TURKI, K I S, and STONE, M. 1978. Petrographic and chemical distinction between the megacrystic members of the Carnmenellis granite, Cornwall. *Proceedings of the Ussher Society*. Vol.4, 182–189.

ANDREWS, J R, BAKER, A J, and PAMPLIN, C F. 1988. A reappraisal of the facing confrontation in north Cornwall: fold- or overthrust-dominated tectonics? *Journal of the Geological Society of London*, Vol. 145, 777–788.

ARKELL, W J. 1943. The Pleistocene rocks at Trebetherick Point, Cornwall; their interpretation and correlation. *Proceedings of the Geologists' Association*, Vol. 54, 141–170.

AUSTIN, D, GERRARD, G A M, and GREEVES, T A P. 1989. Tin and agriculture in the Middle Ages and beyond: landscape archaeology in St Neot Parish, Cornwall. *Cornish Archaeology*, Vol. 28, 5–251.

AUSTIN, R L, DREESDEN, R, SELWOOD, E B, and THOMAS, J M. 1992. New conodont information relating to the Devonian stratigraphy of the Trevone Basin, north Cornwall, south-west England. *Proceedings of the Ussher Society*. Vol. 8, 23–28.

— ORCHARD, M J, and STEWART I J. 1985. Conodonts of the Devonian system from Great Britain. 93–166 in *Stratigraphical index of conodonts*. HIGGINS, A C, and AUSTIN, R L (editors). (Ellis Horwood, British Micropalaeontological Society.)

BARTON, R M. 1966. *A history of the Cornish china clay industry*. (Truro: D Bradford Barton.)

BATSTONE, A E. 1959. The structure and tectonic history of the Tintagel–Davidstow area. *Transactions of the Royal Geological Society of Cornwall*, Vol. 19, 17–32.

BEESE, A P. 1982. The argillite facies of the Middle Devonian succession in north Cornwall. *Proceedings of the Ussher Society*. Vol. 5, 321–32.

— 1984. Stratigraphy of the Upper Devonian argillite succession in north Cornwall. *Geological Magazine*, Vol. 121, 61–9.

BOTT, M H P, DAY, A A, and MASSON-SMITH, D. 1958. The geological interpretation of gravity and magnetic surveys in Devon and Cornwall. *Philosophical Transactions of the Royal Society of London*, Series A, Vol. 251, 161–191.

— HOLDER, A P, LONG, R E, and LUCAS, A L. 1970. Crustal structure beneath the granites of southwest England. 93–102 *in* Mechanism of igneous intrusion. NEWALL, G, and RAST, N (editors). *Geological Journal Special Issue*, No. 2.

BRAMMALL, A. 1926. The Dartmoor granite. *Proceedings of the Geologists' Association*, Vol. 37, 26–38.

— and HARWOOD, H F. 1932. The Dartmoor granites: their genetic relationships. *Quarterly Journal of the Geological Society of London*, Vol. 88, 171–237.

BRAZIER, S, ROBINSON, D, and MATTHEWS, S C. 1979. Studies of illite crystallinity in south-west England: some preliminary results and their geological setting. *Neues Jahrbuch fur Geologie und Palaontologie*, Vol. 11, 641–662.

BRISTOW, C M. 1990. The Fal valley lineaments. *Journal of the Camborne School of Mines*, 1989, 34–41.

— and EXLEY, C S. 1994. Historical and geological aspects of the China Clay industry in south-west England. *Transactions of the Royal Geological Society of Cornwall*, Vol. 21, 247–314

BROMLEY, A V. 1989. Field guide—the Cornubian orefield. *6th International Symposium on Water–Rock Interaction, Malvern (UK)*. International Association of Geochemistry and Cosmochemistry. (Camborne: School of Mines.)

— and HOLL, J. 1986. Tin mineralisation in southwest England. 195–262 in *Mineral processing at a crossroads*. WILLS, B A, and BURLEY, R W (editors). NATO ASI Series, Series E, Applied Sciences No. 117. (Dordrecht: Martinus Nijhoff.)

BROOKS, M, MECHIE, J, and LLEWELLYN, D J. 1983. Geophysical investigations into the Variscides of southwest Britain. 186–197 in *The Variscan Fold Belt in the British Isles*. HANCOCK, P L (editor). (Bristol: Adam Hilger.)

BROWN, A P. 1977. Late-Devensian and Flandrian vegetational history of Bodmin Moor, Cornwall. *Philosophical Transactions of the Royal Society, London*. Series B 276, 251–320.

BURGESS, W G, EDMUNDS, W M, ANDREWS, J N, KAY, R L F, and LEE, D J. 1982. *The origin and circulation of groundwater in the Carnmenellis granite: the hydrochemical evidence. Investigation of the geothermal potential of the UK*. (London: Institute of Geological Sciences.)

BURTON, C J, and TANNER, P W G. 1986. The stratigraphy and structure of the Devonian rocks around Liskeard, east Cornwall, with regional implications. *Journal of the Geological Society of London*, Vol. 143, 96–105.

CASELDINE, C J. 1980. Environmental change in Cornwall during the last 13 000 years. *Cornish Archaeology*, Vol. 19, 3–16.

CHAPPELL, B W, and WHITE, A J R. 1974. Two contrasting granite types. *Pacific Geology*, Vol. 8, 173–174.

CHAROY, B. 1979. Definition et importance des phenomenes deuteriques et des fluides associes dans les granites. Consequences metallogeniques. *Sciences de la Terre, L'Institut National Polytechnique de Lorraine, Nancy, France, Memoir*, No. 37.

— 1982. Tourmalinisation in Cornwall, England. 63–70 in *Metallisation associated with acid magmatism*. EVANS, A M (editor). (Chichester: John Wiley & Sons.)

— 1986. The genesis of the Cornubian batholith (South-West England): the example of the Carnmenellis pluton. *Journal of Petrology*, Vol. 27, 571–604.

CHAYES, F. 1956. *Petrographic modal analysis*. (New York: J. Wiley and Sons.)

CHO, M, and LIOU, J G. 1987. Prehnite-pumpellyite to greenschist facies transition in the Karmutsen metabasalts, Vancouver Island, British Columbia. *Journal of Petrology,* Vol. 28, 417–443.

CLARKE, B B. 1963. Erosional and depositional features of the Camel estuary as evidence of former Pleistocene and Holocene strandlines. *Proceedings of the Ussher Society,* Vol. 1, 57–59.

— 1967. The Sandrock and other features of the cliffs at Mother Ivey's Bay near Padstow. *Proceedings of the Ussher Society,* Vol. 1, 286–287.

— 1971. The Quaternary section at Porth Mear Cove (Abstract). *Proceedings of the Ussher Society,* Vol. 2, 298.

— 1973. The Camel Estuary Pleistocene section west of Tregunna House. *Proceedings of the Ussher Society,* Vol. 2, 551–553.

— 1980. Geomorphology of the Camel valley and estuary. *Proceedings of the Ussher Society,* Vol. 5, 93.

CLAYTON, R E, SCRIVENER, R C, and STANLEY, C J. 1990. Mineralogical and preliminary fluid inclusion studies of lead–antimony mineralisation in north Cornwall. *Proceedings of the Ussher Society,* Vol. 7, 258–262.

CONOLLY, A P, GODWIN, H, and MEGAW, E M. 1955. Studies in the post-glacial history of British vegetation. 11. Late-glacial deposits in Cornwall. *Philosophical Transactions of the Royal Society, London,* Series B 234, 397–469.

DANGERFIELD, J, and HAWKES, J R. 1981. The Variscan granites of SW England: additional information. *Proceedings of the Ussher Society,* Vol. 5, 116–120.

DARBYSHIRE, D P F, and SHEPHERD, T J. 1985. Chronology of granite magmatism and associated mineralisation, SW England. *Journal of the Geological Society of London,* Vol. 142, 1159–1177.

— — 1987. Chronology of magmatism in south-west England: the minor intrusions. *Proceedings of the Ussher Society,* Vol. 6, 431–438.

DAVIS, J C. 1973. Statistics and data analysis in geology. (New York: John Wiley and Sons.)

DE LA BECHE, H T. 1839. Report on the geology of Cornwall, Devon and west Somerset. *Memoir of the Geological Survey of Great Britain.*

DEARMAN, W R. 1963. Wrench faulting in Cornwall and south Devon. *Proceedings of the Geologists' Association,* Vol. 74, 267–287.

DEWEY, H. 1909. On overthrusts at Tintagel (North Cornwall). *Quarterly Journal of the Geological Society of London,* Vol. 65, 265–280.

— 1914. The geology of north Cornwall. *Proceedings of the Geologists' Association,* Vol. 25, 154–179.

— 1915. On spilosites and adinoles from north Cornwall. *Transactions of the Royal Geological Society of Cornwall,* Vol. 15, 71–84.

— 1920. Arsenic and antimony ores. *Memoirs of the Geological Survey, Mineral Resources,* No. 15.

— and FLETT, J S. 1911. On some British pillow lavas and the rocks associated with them. *Geological Magazine,* Vol. 48, 202–209 and 241–248.

DINES, H G. 1956. The metalliferous mining region of South-West England. *Economic Memoir of the Geological Survey of Great Britain.*

DONALD, M B. 1950. Burchard Kranich (c. 1515–1578), miner and Queen's physician, Cornish mining stamps, antimony and Frobishers gold. *Annals of Science,* Vol. 6, 308–322.

DUFFIELD, J, and GILMORE, G R. 1979. An optimum method for the determination of rare-earth-elements by neutron activation analysis. *Journal of Radioanalytical Chemistry,* Vol. 48, 135–145.

DURNING, B D. 1989a. The geological development of the Padstow area, N. Cornwall. Unpublished PhD thesis, University of Exeter.

— 1989b. A new model for the development of the Variscan facing confrontation at Padstow, north Cornwall. *Proceedings of the Ussher Society,* Vol. 7, 141–145.

DURRANCE, E M, and HEALTH, M J. 1985. Thermal groundwater movement and radionuclide transport in SW England. *Mineralogical Magazine,* Vol. 49, 289–299.

EDMONDSON, K M. 1970. A study of the alkali feldspars from some SW England granites. Unpublished MSc thesis, University of Keele.

— 1972. Some aspects of the chemistry of the Bodmin Moor granite. Unpublished PhD thesis, University of Keele.

EVANS, K M 1981. A marine fauna from the Dartmouth Beds (Lower Devonian) of Cornwall. *Geological Magazine,* Vol. 118, 517–523.

— 1983. The marine Lower Devonian of the Plymouth area (abstract). *Proceedings of the Ussher Society,* Vol. 5, 489.

— 1985. The brachiopod faunas of the Meadfoot Group (Lower Devonian) of the Torbay area, south Devon. *Geological Journal,* Vol. 2, 81–90.

EVERARD, C E, LAWRENCE, R H, WITHERICK, M E, and WRIGHT, L W. 1964. Raised beaches and marine geomorphology. 283–310 in *Present views of some aspects of the geology of Cornwall and Devon.* HOSKING, K F G, and SHRIMPTON, G J (editors). 150th Anniversary Volume. (Penzance: Royal Geological Society of Cornwall.)

EXLEY, C S. 1961. A note on greisening in the Bodmin Moor granite. *Geological Magazine,* Vol. 98, 427–430.

— 1965. Some structural features of the Bodmin Moor granite mass. *Proceedings of the Ussher Society,* Vol. 1, 157–159.

— and STONE, M. 1964. The granite rocks of south west England. 131–184 in *Present views on some aspects of the geology of Cornwall and Devon.* HOSKING, K F G, and SHRIMPTON, G J (editors). 150th Anniversary Volume. (Penzance: Royal Geological Society of Cornwall.)

— — and FLOYD, P A. 1983. Composition and petrogenesis of the Cornubian batholith and post-orogenic rocks in SW England. 153–177 In *The Variscan Fold Belt in the British Isles.* HANCOCK, P L (editor). (Bristol: Adam Hilger.)

FEHN, U. 1985. Post–magmatic convection related to high heat production in granites of southwest England: a theoretical study. 99–112 in *High heat production (HHP) granites, hydrothermal circulation and ore genesis.* HALLS, C (editor). Institution of Mining and Metallurgy, Special Publication, St Austell, England.

FERGUSON, C C, and LLOYD, G E. 1982. Palaeostress and strain estimates from boudinage structure and their bearing on the evolution of a major Variscan thrust-fold complex in south-west England. *Tectonophysics,* Vol. 88, 269–289.

FINDLAY, D C, COLBORNE, G J N, COPE, D W, HARROD, T R, HOGAN, D V, and STAINES, S J. 1984. Soils and their use in South West England. *Soil Survey of England and Wales, Bulletin,* No. 14.

FLOYD, P A. 1982. Chemical variation in Hercynian basalts relative to plate tectonics. *Journal of the Geological Society of London,* Vol. 139, 505–520.

— 1984.   Geochemical characteristics and comparison of the Lizard complex and the basaltic lavas within the Hercynian troughs of SW England.   *Journal of the Geological Society of London*, Vol. 141, 61–70.

— EXLEY, C S, and STONE, M. 1983.   Variscan magmatism in southwest England—discussion and synthesis. 178–185 in *The Variscan Fold Belt in the British Isles.* HANCOCK, P L (editor). (Bristol: Adam Hilger.)

— ROWBOTHAM, G. 1979.   Chemical composition of relict clinopyroxenes from the Mullion Island lavas, Cornwall. *Proceedings of the Ussher Society*, Vol. 4, 419–429.

— — 1982.   Chemistry of primary minerals in titaniferous brown amphibole-bearing greenstones from north Cornwall. *Proceedings of the Ussher Society*, Vol. 5, 296–303.

FOOKES, P G, DEARMAN, W R, and FRANKLIN, J A. 1971.   Some engineering aspects of rock weathering with field examples from Dartmoor and elsewhere.   *Quarterly Journal of Engineering Geology*, Vol. 4, 139–185.

FOX, H. 1895.   On a soda feldspar rock at Dinas Head, north coast of Cornwall.   *Geological Magazine*, Vol. 2, 13–20.

— 1905.   Further notes on the Devonian rocks and fossils in the parish of St Minver.   *Transactions of the Royal Geological Society of Cornwall*, Vol. 13, 33–87.

FRANCIS, E H. 1970.   Review of Carboniferous volcanism in England and Wales.   *Journal of Earth Science, University of Leeds*, Vol. 8, 41–56.

FRESHNEY, E C McKEOWN, M C, and WILLIAMS, M. 1972. Geology of the coast between Tintagel and Bude.   *Memoir of the Geological Survey of Great Britain*, Sheet 332.

GAUSS, G A. 1966.   Some aspects of slaty cleavage in the Padstow area of north Cornwall.   *Proceedings of the Ussher Society*, Vol. 1, 221–224.

— 1967.   Structural aspects of the Padstow area, north Cornwall.   *Proceedings of the Ussher Society*, Vol.1, 284–285.

— 1973.   The structure of the Padstow area, North Cornwall.   *Proceedings of the Geologists' Association*, Vol. 84, 283–313.

— and HOUSE, M R. 1972.   The Devonian successions in the Padstow area, North Cornwall.   *Journal of the Geological Society of London*, Vol. 128, 151–72.

GHOSH, P K. 1927.   Petrology of the Bodmin Moor granite (eastern part), Cornwall.   *Mineralogical Magazine*, Vol. 21, 285–309.

— 1934.   The Carnmenellis granite: its petrology, metamorphism and tectonics.   *Quarterly Journal of the Geological Society of London*, Vol. 90, 240–276.

GOODAY, A J. 1973.   Taxonomic and stratigraphic studies on Upper Devonian and Lower Carboniferous Entomozoidae and Rhomboentomozoidae (Ostracoda, Myodocopida) from Southwest England.   Unpublished PhD thesis, University of Exeter.

GOODE, A J J. 1973.   The mode of intrusion of Cornish elvans. *Report of the Institute of Geological Sciences*, No. 73/7.

— HAWKES, J R, DANGERFIELD, J, BURLEY, A J, TAYLOR, R T, and WILSON, A C. 1987.   The geology, petrology and geophysics of the Land's End, Tregonning–Gedolphin and St Michael's Mount Variscan granites.   *Open File Report, British Geological Survey.*

— and TAYLOR, R T. 1988.   Geology of the country around Penzance.   *Memoir of the British Geological Survey.* Sheet 351 and 358 (England and Wales).

HALL, A. 1988.   The distribution of ammonium in granites from South-West England.   *Journal of the Geological Society of London*, Vol. 145, 37–41.

HALLIDAY, A N. 1980.   The timing of early and main stage ore mineralisation in South West Cornwall.   *Economic Geology*, Vol. 75, 752–759.

HAMILTON JENKIN, A K. 1966.   *Mines and miners of Cornwall. XII Around Liskeard.*   62 pp. (Truro: Truro Bookshop.)

— 1970.   *Mines and miners of Cornwall, 16, Wadebridge, Camelford and Bude.*   (Truro: Federation of Old Cornwall Societies.)

HAWKES, J R. 1981.   A tectonic 'watershed' of fundamental consequence in the post–Westphalian evolution of Cornubia. *Proceedings of the Ussher Society*, Vol. 5, 128–131.

— and DANGERFIELD, J. 1978.   The Variscan granites of south–west England: a progress report.   *Proceedings of the Ussher Society*, Vol. 4, 158–171.

— HARDING, R R, and DARBYSHIRE, D P F. 1975.   Petrology and Rb: Sr age of the Brannel, South Crofty and Wherry elvan dykes.   *Bulletin of the Geological Survey of Great Britain*, Vol. 52, 27–42.

HENLY, S. 1974.   Geochemistry and petrogenesis of elvan dykes in the Perranporth area, Cornwall.   *Proceedings of the Ussher Society*, Vol. 3, 128–136.

HOBSON, D M. 1975 (for 1974).   The stratigraphy and structure of the Port Isaac area, North Cornwall.   *Proceedings of the Geologists' Association*, Vol. 85, 473–491.

— and SANDERSON, D J. 1975.   Major folds from the southern margin of the Culm Synclinorium.   *Journal of the Geological Society of London*, Vol. 131, 337–352.

— — 1983.   Variscan deformation in south-west England. 108–129 in *The Variscan Fold Belt in the British Isles.* HANCOCK, P L (editor). (Bristol: Hilger.)

HOGG, T. 1825.   *Manual of mineralogy.*   (Truro.)

HOLDER, A P, and BOTT, M H P. 1971.   Crustal structure in the vicinity of SW England.   *Geophysical Journal of the Royal Astronomical Society*, Vol. 23, 465–489.

HOLDER, M T, and LEVERIDGE, B E. 1986.   A model for the tectonic evolution of south Cornwall.   *Journal of the Geological Society of London*, Vol. 143, 125–134.

HOUSE, M R. 1956.   Devonian goniatites from north Cornwall. *Geological Magazine*, Vol. 93, 257–262.

— 1961.   The Devonian successions of the Padstow area. *Abstract of the Proceedings of the 3rd Conference on the geology and geomorphology of SW England*, 4–5.

— 1963.   Devonian ammonoid successions and facies in Devon and Cornwall.   *Quarterly Journal of the Geological Society of London*, Vol. 119, 1–27.

— and DINELY, D L. 1985.   Devonian Series boundaries in Britain.   *Courier Forschungsinstitut Senckenberg*, Vol. 75, 301–309.

— MOURAVIEFF, N, and BEESE, A P. 1978.   North Cornwall. 57–68 in *A field guide to selected areas of the Devonian of South-west England, International Symposium on the Devonian System.* SCRUTTON, C T (editor). (London: Palaeontological Association.)

— RICHARDSON, J B, CHALONER, W G, ALLEN, J R L, HOOLAND, C H, and WESTOLL, T S. 1977.   A correlation of Devonian rocks of the British Isles.   *Geological Society of London, Special Report*, Vol. 7.

— and SELWOOD, E. B. 1966.   Palaeozoic palaeontology in Devon and Cornwall.   48–86 in *Present views on some aspects of the geology of Cornwall and Devon.* HOSKING, K F G, and SHRIMPTON, G J (editor). 150th Anniversary Volume. (Penzance: Royal Geological Society of Cornwall.)

HULSEMANN, J, and EMERY, K O. 1961.   Stratification in Recent sediments of Santa Barbara basin as controlled by organisms and water character.   *Journal of Geology*, Vol. 69, 279–290.

HUMPHRIES, B, and SMITH, S A. 1989.   The distribution and significance of sedimentary apatite in Lower to Middle Devonian sediments east of Plymouth Sound.   *Proceedings of the Ussher Society*, Vol. 7, 118–124.

ISAAC, K P, TURNER, P J, and STEWART, I J. 1982.   The evolution of the Hercynides of central SW England.   *Journal of the Geological Society of London*, Vol. 139, 521–531.

JACKSON, N J, HALLIDAY, A N, SHEPPARD, S M F, and MITCHELL, J G. 1982.   Hydrothermal activity in the St Just mining district, Cornwall, England.   137–179 in *Metallisation associated with acid magnetism*. EVANS, A M (editor). (Chichester: J Wiley and Sons.)

— WILLIS-RICHARDS, J, MANNING, D A C, and SAMS, M S. 1989. Evolution of the Cornubian ore field, Southwest England: Part II. Mineral deposits and ore-forming processes.   *Economic Geology*, Vol. 84, 1101–1133.

JEFFERIES, N L. 1985.   The origin of sillimanite-bearing pelitic xenoliths within the Carnmenellis pluton, Cornwall. *Proceedings of the Ussher Society*, Vol. 6, 229–236.

— 1988.   The distribution of uranium in the Carnmenellis pluton, Cornwall.   Unpublished PhD thesis, University of Exeter.

KIRCHGASSER, W T. 1970.   Conodonts from near the Middle/Upper Devonian boundary in North Cornwall. *Palaeontology*, Vol. 13, 335–354.

KISCH, H J. 1980.   Incipient metamorphism of Cambro-Silurian clastic rocks from Jamtland Supergroup, central Scandinavian Caledonides, western Sweden: illite crystalinity and "vitrinite" reflectance.   *Journal of the Geological Society of London*, Vol. 137, 271–288.

LEAT, P T, THOMPSON, R N, MORRISON, M A, HENDRY, G L, and TRAYHORN, S C. 1987 (for 1986).   Geodynamic significance of post-Variscan intrusive and extrusive potassic magmatism in SW England.   *Transactions of the Royal Society of Edinburgh: Earth Sciences*, Vol. 77, 349–360.

LEE, M K, BROWN, G C, WEBB, P C, WHEILDON, J, and ROLIN, K E. 1987.   Heatflow, heat production and thermo-tectonic setting in mainland UK.   *Journal of the Geological Society of London*, Vol. 144, 35–42.

LEVERIDGE, B E, HOLDER, M T, and GOODE, A J J. 1990.   Geology of the country around Falmouth.   *Memoir of the British Geological Survey*, Sheet 352 (England and Wales.)

LISTER, C J. 1984.   Xenolith assimilation in the granites of south–west England.   *Proceedings of the Ussher Society*, Vol. 6, 46–53.

MANNING, D A C. (1991).   Chemical variation in tourmalines from south-west England.   *Proceedings of the Ussher Society*. Vol. 7, 327–332.

MCKEOWN, M C, EDMUNDS, A E, WILLIAMS, M, FRESHNEY, E C, and MASSON SMITH, D J. 1973.   Geology of the Country around Boscastle and Holsworthy.   *Memoir of the Geological Survey of Great Britain*, Sheets 322 and 323.

MEREFIELD, J R. 1989.   Organic protection of aragonite in Recent dune sands.   *Proceedings of the Ussher Society*, Vol. 7, 185–187.

MOURAVIEFF, N A. 1977.   Additional conodonts from near the Middle/Upper Devonian boundary in North Cornwall. *Palaeontology*, Vol. 13, 335–354.

OLIVER, W A and CHLUPAC, I. 1991.   Defining the Devonian. *Lethaia*, Vol. 24, 119–122.

PALMER, J, and NEILSON, R A. 1962.   The origin of granite tors on Dartmoor.   *Proceedings of the Yorkshire Geological Society*. Vol. 33, 315–340.

PAMPLIN, C F. 1990.   A model for the tectono-thermal evolution of north Cornwall.   *Proceedings of the Ussher Society*, Vol. 7, 206–211.

— and ANDREWS, J R. 1988.   Timing and sense of shear in the Padstow Facing Confrontation.   *Proceedings of the Ussher Society*, Vol. 7, 73–76.

PARKINSON, J. 1903.   The geology of the Tintagel and Davidstow district.   *Quarterly Journal of the Geological Society of London*, Vol. 59, 408–421.

PASCOE, W A. 1945.   The Foweymoor district.   *Mining Magazine*, Vol. 72, 210–212.

PATTISON, S R. 1847.   On some post–Tertiary deposits in Cornwall.   *Transactions of the Royal Geological Society of Cornwall*. Vol. 7, 34–36.

PICHAVANT, M. 1979.   Étude experimentale a haute temperature et 1 kbar due rôle du bore dans quelques systèmes silicates. Intérêt pétrologique et métallogenique.   Thèse speciale. L'Institut National Polytechnique de Lorraine, Nancy, France.

POUND, C J. 1983.   The sedimentology of the Lower–Middle Devonian Staddon Grits and Jennycliff Slates on the east side of Plymouth Sound.   *Proceedings of the Ussher Society*, Vol. 5, 465–472.

POWER, G M. 1968.   Chemical variation in tourmalines from South-west England.   *Mineralogical Magazine*, Vol. 36, 1078–1089.

PRIMMER, T J. 1985a.   A transition from diagenesis to green–schist facies within a major Variscan fold/thrust complex in south-west England.   *Mineralogical Magazine*, Vol. 49, 365–374.

— 1985b.   The pressure–temperature history of the Tintagel district, Cornwall: metamorphic evidence on the tectonic evolution of the area.   *Proceedings of the Ussher Society*, Vol. 6, 218–223.

PRYCE, W. 1778.   *Mineralogia Cornubiensis: a treatise on minerals, mines and mining*. (London.)

RATTEY P R, and SANDERSON, D J. 1982.   Patterns of folding within nappes and thrust sheets: examples from the Variscan of Southwest England.   *Tectonophysics*, Vol. 88, 247–67.

REID, C, BARROW, G, and DEWEY, H. 1910.   The geology of the country around Padstow and Camelford. Explanation of sheets 335 and 336.   *Memoir of the Geological Survey of Great Britain*.

— and DEWEY, H. 1908.   The origin of the pillow lavas near Port Isaac in Cornwall.   *Quarterly Journal of the Geological Society of London*, Vol. 64, 264–272.

RICE-BIRCHALL, B. 1991.   Petrology and geochemistry of basic volcanics, N. Cornwall.   Unpublished PhD thesis, University of Keele.

— and FLOYD, P A. 1988.   Geochemical and source characteristics of the Tintagel Volcanic Formation.   *Proceedings of the Ussher Society*, Vol. 7, 52–55.

RIPLEY, M J. 1964.   The geology of the coastal area in the vicinity of Newquay, Cornwall.   Unpublished PhD thesis, University of Birmingham.

ROBERTS, R L, and SANDERSON, D J. 1971. Polyphase development of slaty cleavage and the confrontation of facing directions in the Devonian rocks of North Cornwall. *Nature, London,* Vol. 230, 87–89.

ROBINSON, D, and READ, D. 1981. Metamorphism and mineral chemistry of green-schists from Trebarwith Strand, north Cornwall. *Proceedings of the Ussher Society,* Vol. 5, 132–38.

— and SEXTON, D. 1987. Geochemistry of the Tintagel Volcanic Formation. *Proceedings of the Ussher Society,* Vol. 6, 523–528.

RUDWICK, M J S. 1985. *The Great Devonian Controversy.* (Chicago and London: University of Chicago Press.)

RUSSELL, A. 1949. *An account of the antimony mines of Great Britain and Ireland and the minerals found therein.* Unpublished manuscript, British Museum (NH) Library.

SAMS, M S, and THOMAS-BETTS, A. 1988. Models of convective fluid flow and mineralisation in south-west England. *Journal of the Geological Society of London.* Vol. 145, 809–917.

SANDERSON, D J. 1979. The transition from upright to recumbent folding in the Variscan fold belt of south-west England: a model based on the kinematics of simple shear. *Journal of Structural Geology,* Vol. 1, 171–180.

SCOURCE, J D. 1985. Late Pleistocene stratigraphy of the Isles of Scilly and adjoining regions. Unpublished PhD thesis University of Cambridge.

SCRIVENER, R C. 1989. Geology and minralogy of West Colliford mill. 199–203 in *Tin and agriculture in the Middle Ages and beyond: landscape archaeology in St Neot Parish, Cornwall.* AUSTIN, D, GERRARD, G A M, and GREEVES, T A P (editors). *Cornish Archaeology,* No 28.

SEAGO, R D, and CHAPMAN, T J. 1988. Confrontation of structural styles and the evolution of a foreland basin in Central SW England. *Journal of the Geological Society of London,* Vol. 145, 789–800.

SEILACHER, A. 1982. Distinctive features of sandy tempestites. 333–349 in *Cyclic and event stratification.* EINSELE, G, and SEILACHER, A (editors). (Berlin: Springer-Verlag.)

SELWOOD, E B. 1990. A review of basin development in central south-west England. *Proceedings of the Ussher Society,* Vol. 7, 199–205.

— EDWARDS, R A, SIMPSON, S, CHESHER, J A, HAMBLIN, R J O, HENSON, M R, RIDDOLLS, B W, and WATERS, R A. 1984. Geology of the country around Newton Abbot. *Memoir of the British Geological Survey,* Sheet 339 (England and Wales.)

— STEWART, I J, and THOMAS, J M. 1985. Upper Palaeozoic sediments and structure in north Cornwall—a reinterpretation. *Proceedings of the Geologists' Association,* Vol. 96, 129–141.

— — 1986a. Upper Palaeozoic successions and nappe structures in North Cornwall. *Journal of the Geological Society of London,* Vol. 143, 75–82.

— — 1986b. Variscan facies and structure in central SW England. *Journal of the Geological Society of London,* Vol. 143 199–207.

— — 1988. The Padstow Confrontation, north Cornwall: a reappraisal. *Journal of the Geological Society of London,* Vol. 139, 533–541.

— — 1993. The Tredorn Nappe, north Cornwall: a review. *Proceedings of the Ussher Society,* Vol. 8, 89–93.

— — BORLEY, G D, and DEAN, A. 1993. A revision of the Upper Palaeozoic stratigraphy of the Trevone Basin, north Cornwall, and its regional significance. *Proceedings of the Geologists' Association,* Vol. 104, 137–148.

SHACKLETON, R M, RIES, A C, and COWARD, M P. 1982. An interpretation of the Variscan structures in SW England. *Journal of the Geological Society of London,* Vol. 139, 533–541.

SHEPPARD, S M F. 1977. The Cornubian batholith, south west England: D/H and 180/160 studies of kaolinite and other alteration minerals. *Journal of the Geological Society of London,* Vol. 133, 573–591.

SHEPHERD, T J, and SCRIVENER, R C. 1987. The role of basinal brines in the genesis of polymetalic vein deposits, Kit Hill–Gunnislake area, SW England. *Proceedings of the Ussher Society,* Vol. 6, 491–497.

SIMPSON, P R, BROWN, G C, PLANT, J A, and OSTLE, D. 1979. Uranium mineralisation and granite magmatism in Britain. *Philosophical Transactions of the Royal Society of London,* Series A, 291, 385–412.

STAINES, S J. 1976. Soils in Cornwall I: Sheet SX18 (Camelford). *Soil Survey Record,* No 34. (Harpenden: Soil Survey.)

STEWART I J. 1981. Late Devonian and Lower Carboniferous conodonts from N. Cornwall and their stratigraphic significance. *Proceedings of the Ussher Society,* Vol. 5, 179–85.

— 1983. The structure, stratigraphy and conodont biostratigraphy of the north eastern margin of Bodmin Moor and adjacent areas. Unpublished PhD thesis, University of Exeter.

STONE, M. 1968. A study of the Praa Sands elvan and its bearing on the origin of elvans. *Proceedings of the Ussher Society,* Vol. 2, 37–42.

— 1988. The significance of almandine garnets in the Lundy and Dartmoor granites. *Mineralogical Magazine,* Vol. 52, 651–658.

— and EXLEY C S. 1978. A cluster analysis of chemical data from the granites of S.W. England. *Proceedings of the Ussher Society,* Vol. 4, 172–18.

— — 1986. High heat production granites of southwest England and their associated mineralisation: a review. *Transactions of the Institution of Mining and Metallurgy,* Section B, Applied Earth Science, Vol. 95, 25–36.

STRECKEISEN, A. 1976. To each plutonic rock its proper name. *Earth Science Reviews,* Vol. 12, 1–33.

SUN, S S, and McDONOUGH, W F. 1989. Chemical and isotopic systematics of oceanic basalts: implications for mantle composition and processes. 313–345 in *Magmatism in the ocean basins.* SAUNDERS, A D, and NORRY, M J (editors). Special Publication of the Geological Society of London, No. 42. (Oxford: Blackwell Scientific Publications.)

TEPUNGA, M T. 1957. Periglaciation in southern England. *Tijdschrift van het Komninklijk Nederlandsh Aardrijkskundig Genootschap,* Vol. 74, 401–12.

THOMAS, I. 1909. Notes on the Trilobite fauna of Devon and Cornwall. *Geological Magazine,* Vol. 56, 193–204.

TOMBS, J M C. 1977. A study of the space form of the Cornubian batholith and its application to detailed gravity surveys in Cornwall. *Mineral Reconnaissance Report, Institute of Geological Sciences,* No. 11.

TURNER, R G. 1968. The influence of granite emplacement on structures in south-west England. Unpublished PhD thesis, University of Newcastle-upon-Tyne.

TUCKER, M E. 1969. Crinoidal turbidites from the Devonian of Cornwall and their palaeogeographic importance. *Sedimentology,* Vol. 13, 281–90.

WALSH, J N. 1982. Whole rock analysis by inductively coupled plasma spectrometry. 79–91 in *Sampling and analysis for the mining industry*. (London: Institution of Mining and Metallurgy.)

WARR, L N. 1988. The deformational history of the area northwest of the Bodmin Moor granite. *Proceedings of the Ussher Society*, Vol. 7, 67–72.

— 1989. The structure and evolution of the Davidstow Anticline, and its relationship to the Southern Culm Overfold, north Cornwall. *Proceedings of the Ussher Society*, Vol. 7, 136–140.

— 1991. Basin inversion in the external Variscan of SW England, UK. Unpublished PhD thesis, University of Exeter.

— 1993. Basin inversion and foreland basin development in the Rhenohercynian of south-west England. 197–224 in *Rhenohercynian and Subvariscan fold belts*. GAYER, R A, GREILING, R O, and VOGEL, A K (editors). (Braunschweig/Wiesbaden: Vieweg Publishing.)

— and DURNING B. 1990. Discussion on: A reappraisal of the facing confrontation in north Cornwall: fold or thrust-dominated tectonics? *Journal of the Geological Society of London*, Vol. 147, 408–510.

— PRIMMER, T J, and ROBINSON, D. 1991. Variscan very low-grade metamorphism in southwest England: a diastathermal and thrust related origin. *Journal of Metamorphic Geology*, Vol. 9, 751–764.

— and ROBINSON, D. 1990. The application of illite "crystallinity" technique to geological interpretation: a case study from north Cornwall. *Proceedings of the Ussher Society*, Vol. 7, 224–228.

WATSON, J, FOWLER, M B, PLANT, J A, and SIMPSON, P R. 1984. Variscan–Caledonian comparisons: late orogenic granites. *Proceedings of the Ussher Society*, Vol. 6, 2–12.

WHEILDON, J. FRANCIS, M F, ELLIS, J R L, and THOMAS-BETTS, A. 1981. Investigation of the south-west England thermal anomaly zone. Report EUR 7276 EN, Directorate-General for Research, Science and Evaluation, Commission of the European Communities. 410 pp.

WHITLEY, N. 1849. On the remains of ancient volcanoes on the north coast of Cornwall. *30th Annual Report of the Royal Institution of Cornwall*, 47.

WILLIAMS, C T, and FLOYD, P A. 1981. The localized distribution of U and other incompatible elements in spilitic pillow lavas. *Contributions to Mineralogy and Petrology*, Vol. 78, 111–117.

WILLIS-RICHARDS, J, and JACKSON, N J. 1989. Evolution of the Cornubian ore field: Part I. Batholith modelling and ore distribution. *Economic Geology*, Vol. 84, 1078–1100.

WILSON, G. 1951. The tectonics of the Tintagel area, North Cornwall. *Quarterly Journal of the Geological Society of London*, Vol. 106, 393–432.

ZIEGLER, W, and SANDBERG, C A. 1990. The Late Devonian standard conodont zonation. 1–115 *in* 1st International Senckenberg Conference and 5th European Conodont Symposium (ECOS V) Contributions V. ZIEGLER, W (editor). *Courier Forschungsinstitut Senckenberg*, Vol. 121.

APPENDIX 1

British Geological Survey, Borehole Records

At the time of going to press, 112 borehole records of the Trevose Head/Camelford district are held in BGS archives, 31 of these are more than 20 m deep. Additional information is constantly being added as it becomes available.

The full records are held in the National Geological Records Centre, BGS, Keyworth. Most are available in full for inspection by prior appointment, and copies of records may be made;. a charge is made for these services, which reflects the storage and maintenance of the archive.

APPENDIX 2

1:10 000 scale maps and reports.

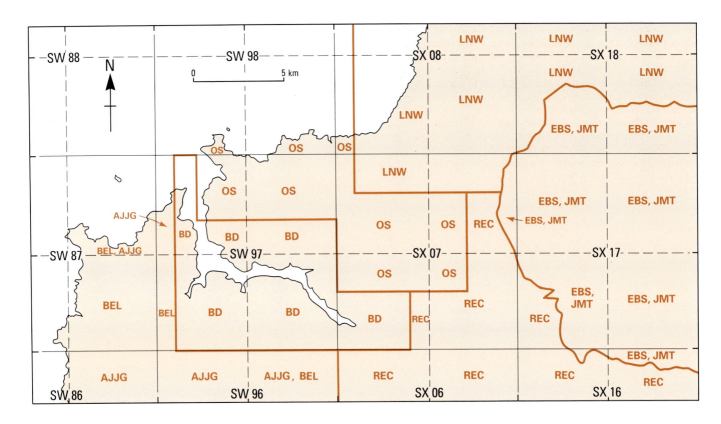

The component 1:10 000 scale National Grid sheets of Geological Sheet 335 and 336 are shown in the diagram above. The surveyors were R E Clayton, B Durning, O Smith, L N Warr, E B Selwood and J M Thomas (University of Exeter): A J J Goode and B E Leveridge (British Geological Survey). Uncoloured dyeline copies of the maps are available for purchase from the British Geological Survey, Keyworth, Nottingham, NG12 5GG.

The areas mapped by each surveyor are indicated by initials and described in six Technical Reports:

Geological notes and details for 1:10 000 sheets SW 86 NW and NE (combined, part), SW 87 NE, SW (part) and SE, SW 96 NW (part) and NE (part), SW 97 NW (part) and SW (part). (Trevose Head and St Breock Downs). A J J Goode and B E Leveridge, 1991.

Geological notes and details for 1:10 000 sheets SW 07 SW (part), SW 97 NW (part), NE (part), SW (part) and SE (Padstow area, north Cornwall district). B Durning, 1989. WA/94/64.

Geological notes and details for 1:10 000 sheets SW 97 NW (part) and NE (part), SW 98 SW, and SE, SX 07 NW (part), NE (part), SW (part), and SE (part), SX 08 SW (part) (St Miniver area, north Cornwall district). O Smith, 1991. WA/94/67.

Geological notes and details for 1:10 000 sheets SX 06 NW (part) and NE (part), SX 07 NE (part), SW (part) and SE (part), SX 16 NW (part) and NE (part), SX 17 SW (part), (Helland area, north Cornwall district). R E Clayton, 1992. WA/94/68.

Geological notes and details for 1:10 000 sheets SX 16 NW (part) and NE (part), SX 07 NE (part), SX 17 SW (part) SE, NW and NE, SX 18 SW (part) and SE (part), (Bodmin Moor area, north Cornwall district). E B Selwood and J M Thomas, 1990. WA/94/66.

Geological notes and details for 1:10 000 sheet SX 07 NW (part) and NE (part), SX 18 SW (part) and SE (part), SX 08 SW (part) and SE. (Camelford area, north Cornwall district). L N Warr, 1990. WA/94/65.

APPENDIX 3

British Geological Survey photographs.

Photographs illustrating the geology and scenery of the district are deposited in the libraries of the British Geological Survey at Keyworth, Nottingham NG12 5GG; in Murchison House, West Mains Road, Edinburgh EH9 3LA; in the regional office, St Just, 30 Pennsylvania Road, Exeter EX4 6BX, and in the BGS Information Office, Natural History Museum Earth Galleries, Exhibition Road, London SW7 2DE.

Some of the photographs, taken in the early part of the century, are of historical interest and show the old tin and copper mines. These are available in black and white. The more recent photographs can be supplied as black and white or colour prints and colour transparencies, at a fixed tariff.

AUTHOR CITATIONS
FOR FOSSIL SPECIES

To satisfy the rules and recommendations of the international codes of botanical and zoological nomenclature, authors of cited species are listed below.

Chapter Two

Camptotriletes paprothii Higgs and Streel, 1984
Cyrtospirifer verneuili (Murchison, 1840)
Dibolisporites albitiensis McGregor and Camfield, 1976
Dictyotriletes emsiensis (Allen) McGregor, 1973
Emphanisporites rotatus (McGregor) McGregor, 1973
Geminospora lemurata (Balme) Playford, 1983
Gorgonisphaeridium condensum Playford, 1981
Lophozonotriletes lebedianensis Naumova, 1953
Patella vulgata Linnaeus, 1738
Ponticeras pedderi House, 1963
Ponticeras prumiense (Steininger, 1849)
Retusotriletes communis Naumova, 1953
Retusotriletes triangulatus (Streel) Streel, 1967
Solisphaeridium inaffectum Playford, 1981
Strophonelloides reversa (Calvin, 1878)
Velamisporites perinatus (Hughes and Playford) Playford, 1971
Veryhachium valiente Cramer, 1964

INDEX

BRITISH GEOLOGICAL SURVEY

Keyworth, Nottingham NG12 5GG
0115 936 3100

Murchison House, West Mains Road, Edinburgh
EH9 3LA   0131-667 1000

London Information Office, Natural History Museum
Earth Galleries, Exhibition Road, London SW7 2DE
0171-589 4090

The full range of Survey publications is available through the
Sales Desks at Keyworth and at Murchison House, Edinburgh,
and in the BGS London Information Office in the Natural
History Museum (Earth Galleries). The adjacent bookshop
stocks the more popular books for sale over the counter. Most
BGS books and reports can be bought from The Stationery
Office and through Stationery Office agents and retailers.
Maps are listed in the BGS Map Catalogue, and can be bought
together with books and reports through BGS-approved
stockists and agents as well as direct from BGS.

*The British Geological Survey carries out the geological survey of Great
Britain and Northern Ireland (the latter as an agency service for the
government of Northern Ireland), and of the surrounding continental
shelf, as well as its basic research projects. It also undertakes
programmes of British technical aid in geology in developing countries
as arranged by the Department for International Development and
other agencies.*

*The British Geological Survey is a component body of the Natural
Environment Research Council.*

Published by The Stationery Office and available from:

The Publications Centre
(mail, telephone and fax orders only)
PO Box 276, London SW8 5DT
General enquiries 0171 873 0011
Telephone orders 0171 873 9090
Fax orders 0171 873 8200

The Stationery Office Bookshops
59–60 Holborn Viaduct, London EC1A 2FD
temporary until mid 1998
(counter service and fax orders only)
Fax 0171 831 1326
68–69 Bull Street, Birmingham B4 6AD
0121 236 9696   Fax 0121 236 9699
33 Wine Street, Bristol BS1 2BQ
0117 9264306 Fax 0117 9294515
9–21 Princess Street, Manchester M60 8AS
0161 834 7201 Fax 0161 833 0634
16 Arthur Street, Belfast BT1 4GD
01232 238451 Fax 01232 235401
The Stationery Office Oriel Bookshop
The Friary, Cardiff CF1 4AA
01222 395548 Fax 01222 384347
71 Lothian Road, Edinburgh EH3 9AZ
(counter service only)

Customers in Scotland may
mail, telephone or fax their orders to:
Scottish Publications Sales
South Gyle Crescent, Edinburgh EH12 9EB
0131 228 4181 Fax 0131 622 7017

The Stationery Office's Accredited Agents
(see Yellow Pages)

and through good booksellers